T0242854

Springer Theses

Recognizing Outstanding Ph.D. Research

Aims and Scope

The series "Springer Theses" brings together a selection of the very best Ph.D. theses from around the world and across the physical sciences. Nominated and endorsed by two recognized specialists, each published volume has been selected for its scientific excellence and the high impact of its contents for the pertinent field of research. For greater accessibility to non-specialists, the published versions include an extended introduction, as well as a foreword by the student's supervisor explaining the special relevance of the work for the field. As a whole, the series will provide a valuable resource both for newcomers to the research fields described, and for other scientists seeking detailed background information on special questions. Finally, it provides an accredited documentation of the valuable contributions made by today's younger generation of scientists.

Theses are accepted into the series by invited nomination only and must fulfill all of the following criteria

- They must be written in good English.
- The topic should fall within the confines of Chemistry, Physics, Earth Sciences, Engineering and related interdisciplinary fields such as Materials, Nanoscience, Chemical Engineering, Complex Systems and Biophysics.
- The work reported in the thesis must represent a significant scientific advance.
- If the thesis includes previously published material, permission to reproduce this must be gained from the respective copyright holder.
- They must have been examined and passed during the 12 months prior to nomination.
- Each thesis should include a foreword by the supervisor outlining the significance of its content.
- The theses should have a clearly defined structure including an introduction accessible to scientists not expert in that particular field.

More information about this series at http://www.springer.com/series/8790

Yuen Wu

Controlled Synthesis of Pt–Ni Bimetallic Catalysts and Study of Their Catalytic Properties

Doctoral Thesis accepted by
Tsinghua University, Beijing, China

 Springer

Author
Dr. Yuen Wu
Department of Chemistry
Tsinghua University
Beijing
China

Supervisor
Prof. Yadong Li
Department of Chemistry
Tsinghua University
Beijing
China

ISSN 2190-5053
Springer Theses
ISBN 978-3-662-57040-1
DOI 10.1007/978-3-662-49847-7

ISSN 2190-5061 (electronic)

ISBN 978-3-662-49847-7   (eBook)

This Springer imprint is published by Springer Nature
The registered company is Springer-Verlag GmbH Berlin Heidelberg

Parts of this thesis have been published in the following journal articles:

1. Wu YE, Wang DS, Zhao P, Niu ZQ, Peng Q, Li YD (2011) Monodispersed Pd–Ni Nanoparticles: Composition Control Synthesis and Catalytic Properties in the Miyaura-Suzuki Reaction. Inorg Chem 50 (6):2046–2048
2. Wu Y, Cai S, Wang D, He W, Li Y (2012) Syntheses of Water-Soluble Octahedral, Truncated Octahedral, and Cubic Pt–Ni Nanocrystals and Their Structure–Activity Study in Model Hydrogenation Reactions. J Am Chem Soc 134 (21):8975–8981
3. Wu Y, Wang D, Niu Z, Chen P, Zhou G, Li Y (2012) A strategy for designing a concave Pt–Ni alloy through controllable chemical etching. Angew Chem Int Ed 51 (50):12524–12528
4. Wu Y, Wang D, Chen X, Zhou G, Yu R, Li Y (2013) Defect-Dominated Shape Recovery of Nanocrystals: A New Strategy for Trimetallic Catalysts. J Am Chem Soc 135 (33):12220–12223
5. Wu Y, Wang D, Zhou G, Yu R, Chen C, Li Y (2014) Sophisticated Construction of Au Islands on Pt–Ni: An Ideal Trimetallic Nanoframe Catalyst. J Am Chem Soc 136 (33):11594–11597
6. Wu Y, Wang D, Li Y (2014) Nanocrystals from solutions: catalysts. Chem Soc Rev 43 (7):2112–2124

Supervisor's Foreword

This thesis summarizes Dr. Yuen Wu's scientific achievements that were made during his doctoral study at the Department of Chemistry of Tsinghua University. As his Ph.D. supervisor, I think Dr. Wu has made important contributions to improve the understanding in principles of nucleation and growth of Pt–Ni bimetallic nanostructures. Among that, I want to introduce the major findings as follows:

The commonly obtained Pt–Ni alloy is usually capped with long-chain capping agents, which will result in low activity and selectivity. He developed a general method for shape-controlled synthesis of water-soluble Pt–Ni alloy. These catalysts are capped with hydrophilic polymer and exhibit excellent catalytic performance towards several hydrogenation reactions in polar solvents.

Chemical etching is usually too drastic to control. He developed a priority-related chemical etching method to transfer the starting Pt–Ni octahedron into concave structure parallel to the original octahedron. The concave Pt–Ni alloy has high density of atomic steps and exhibits great potential as a new type of material for fuel cell, organic catalysis, and so on.

Considering different nucleation and growth rates during the synthesis of nanoparticles, traditional synthetic methodology such as co-reduction and thermal decomposition are becoming insufficient to completely control the trimetallic nanostructures. Combining the galvanic replacement and chemical etching, he developed a sequential method to synthesize trimetallic nanoframe. In addition, he discovered a novel shape recovery of trimetallic nanocrystals which was dominated by the surface defects. On the basis of the rendered "step-induced/terrace-assisted" growth mechanism, it can be confirmed unequivocally two principles allowing for the control over the shape and segregation in trimetallic system.

Beijing, China   Prof. Yadong Li
March 2016

Acknowledgments

I would like to express my gratitude to Prof. Yadong Li, not only because of his inestimable comments and suggestions, but also his meaningful advice and great encouragement throughout my Ph.D. study. I learned how to summarize the intrinsic principles of physics and chemistry. Over the past 5 years, he taught me how to handle the relationship between life and science. Also, I learned how to be nice with everybody, which is very crucial for my works. He spent a significant amount of time and energy to care about me. He did an excellent job as a leader to guide me to the World of Science. Without his guidance and stimulation, I may have missed my direction during my studying way.

I would like to thank Prof. Qing Peng especially for providing useful advice with regard to my experimental details and for his kind regards to my life. He is getting very well with me due to his amiable personality. He did give a lot of help to me when I served as the scientific assistant in the lab.

In addition, I would like to acknowledge Prof. Dingsheng Wang and Prof. Gang Zhou for their kindly help and advices. They are more likely my friends than teacher. They did great help to my preparation of my first manuscript, which is of great help for my future research. They also kindly discuss with me about my initial idea about the synthesis of Pt–Ni nanostructures.

Finally, I would like to express my deep gratitude to my parents, my wife, and my friends that concern with me very much for their support and warm encouragement. I also want to thank the financial support from the NSFC.

Contents

Chapter 1
Introduction

1.1 Background of Research

The ultimate goal of nanoscale science and technology is to manipulate single atoms, assemble atoms in a controllable way, and design nanostructured materials with desired physical and chemical properties. In order to achieve this goal, the nucleation and growth mechanism of nanocrystals (NCs) as well as the relationship between the macroscopic properties and microscopic structures of nanocrystals should be fully understood. Because of a high surface area and specific quantum size effect, nanomaterials have drawn widespread concerns in catalysis science (Fig. 1.1).

By greatly expanding the surface area and overcoming the drawback of recovery, nanocrystal catalysis has gradually become a hot research field in recent years. The outstanding physical and chemical properties of nanomaterials impel catalysts to be widely used in fields of chemical engineering, energy conversion, and medicine. As one of the most important materials, metal catalysts especially the noble metals (e.g., Au, Pd, Pt etc.) have become indispensable in most of the catalytic processes. With the technological development of the synthesis and characterization in nanoscience, it has been realized that the activity, selectivity, and stability are closely related to the geometric structure and electronic structure of nanocatalyst. Therefore, designing efficient nanocatalyst with excellent activity, selectivity, and stability at atomic scale exhibit scientific research significance and the long-term prospects for practical applications.

We should first understand the synthetic process of nanocrystals. Nucleation (from precursor to nucleus) and growth (from nucleus to nanocrystal) are two distinct but sequential stages during the preparation of nanocrystals in the solution. A major challenge is to understand the difference between homogeneous and heterogeneous catalyses in the nanocrystalline catalytic reaction, which is considered to correspond to the process from nucleation (homogeneous) to

© Springer-Verlag Berlin Heidelberg 2016
Y. Wu, *Controlled Synthesis of Pt–Ni Bimetallic Catalysts and Study of Their Catalytic Properties*, Springer Theses,
DOI 10.1007/978-3-662-49847-7_1

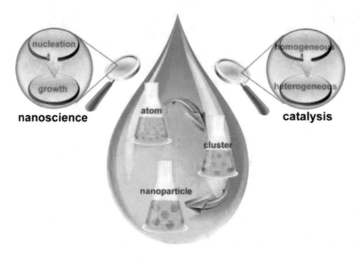

Fig. 1.1 The schematic diagram of the nucleation and growth of nanocrystals

growth (heterogeneous) in nanocrystal synthesis. Understanding the nucleation and growth mechanism of nanocrystals has great significance for designing a nanocrystal catalyst and elucidating the problems in homogeneous and heterogeneous catalyses.

1.1.1 Synthesis of Nanocrystalline Catalyst

On December 29, 1959, Richard P. Feynman stated at the annual meeting of the American Physical Society that "ultimately—in the great future—we can arrange the atoms the way we want; the very atoms, all the way down! What would happen if we could arrange the atoms one by one the way we want them." This classic talk initiated the research of nanoscience and nanotechnology. Decades later, rearranging the atoms in the gas phase had really been achieved by scientists. In 1990, Eigler and Schweizer reported the use of the scanning tunneling microscope (STM) in ultrahigh vacuum to position individual xenon atoms on a single-crystal nickel surface with atomic precision [1]. However, so far, manipulating single atoms in solutions and achieving full control over them are still challenging.

Although rearranging the atoms in the solution phase with atomic precision could not be achieved at present, controllable synthesis of all kinds of nanomaterials with uniform size, well-defined morphology, and unique structure has been greatly developed in recent decades [2]. Based on the experimental experiences and theoretical calculations, the nucleation and growth process of nanomaterials can be well controlled by tuning the thermodynamic and kinetic parameters of the synthetic system. However, the nucleation and growth mechanism of nanocrystals (especially the nucleation stage) could not yet be fully understood, which is indispensable to achieve the goal of manipulating single atoms in solutions.

On the other hand, the underlying reason for the significance of nanostructured materials is their novel catalytic properties when the size was reduced to nanoscale. A classic example is that, although Au in the bulk state is always very unreactive, nanoscale Au particles (smaller than 10 nm) were proved to be catalytically active for low-temperature CO oxidation. Generally speaking, the reason for the superior catalytic properties of nanocrystals is their small particle size and the resulting large surface area. But, in fact, this is not necessary in some case. In 1998, Goodman and coworkers proved that Au nanocrystals with size of 2.5 ~ 3.5 nm showed high reactivity for low-temperature CO oxidation and particles with smaller size (<2 nm) exhibited low reactivity although they possessed a larger surface area. In 2005, our group demonstrated that CeO_2 nanorods exposing reactive {001} and {110} planes generated better catalytic performance for CO oxidation compared to CeO_2 nanocrystals with a smaller particle size and a higher surface area which predominantly exposed stable {111} planes. All of these examples indicate that the catalytic properties of nanocrystals are dependent on various structural factors such as size, composition, and surface environment. Actually, since the discovery of high catalytic activity of Au nanocrystals, the essence of nanocatalysis and the relationship between microstructures of nanocrystals and their catalytic performance have been widely investigated. However, to date, it is still difficult to deeply understand the nanocatalytic process at the atomic level.

First, we reviewed the latest research developments and experimental methods to exploit the nucleation and growth process of nanocrystals. Indeed, during the past decades, incredible success in controllable synthesis of nanocrystals has been achieved. Various synthetic strategies for different kinds of nanomaterials including metals, metal oxides, semiconductors, fluorides, etc., have been developed. Table 1.1 summarizes the main solution-based methods for nanocrystal synthesis.

In 2007, our group presented a review on the interface-mediated growth of monodispersed nanocrystals [3]. Via careful design of the nanoscale interfaces, we could prepare a series of novel nanocrystals. In 2010, Modeshia and Walton summarized the recent efforts in the synthesis of oxide solids using solvothermal routes [4]. Solvothermal/hydrothermal methods are also effective for preparing other nanomaterials. In 2011, Varma et al. gave an overview of producing silver nanostructures by a microwave-assisted synthetic approach which has been also applied to prepare various oxides and chalcogenides [5]. In 2012, Sui and Charpentier provided a comprehensive review article introducing synthesis of metal oxide nanostructures by sol-gel chemistry [6]. Very recently, Suslick et al. reviewed the sonochemical synthesis of nanomaterials [11]. Yin et al. reviewed the templated synthesis of nanostructured materials [12]. Huang et al. reviewed the biomimetic synthesis of nanocrystals [13]. Zheng et al. discussed the roles of small adsorbates in shape control of Pd and Pt nanocrystals [10].

Table 1.2 summarizes recent reviews focusing on various kinds of nanomaterials including metals [22], metal oxides [15], metal chalcogenides [23], and lanthanide-doped nanocrystals [17], and Table 1.3 summarizes recent reviews focusing

Table 1.1 Summaries of the main solution-based methods for nanocrystal synthesis

Synthetic strategies	Characteristics and advantages	References
Interface-mediated synthesis	Nanocrystals can be readily synthesized by rationally tuning the chemical reactions at various interfaces including gas–liquid, liquid–solid, and organic–inorganic interfaces	[3]
Hydrothermal and solvothermal methods	They allow preparation of nanocrystals at temperatures below those required by traditional solid-state reactions via increasing the solubility and reactivity of precursors	[4]
Microwave-assisted synthesis	It allows greener synthesis of nanocrystals with advantages such as shorter reaction times, reduced energy consumption, and better product yields	[5]
Sol–gel synthetic method	It is considered as "soft chemistry" with advantages including high yield, low operation temperatures, and low production costs	[6]
Sonochemical synthesis	Sonochemistry produces unique hot spots that can achieve high temperatures and pressures, and high heating and cooling rates	[7]
Templated synthesis	It uses a preexisting template with desired nanoscale features to direct the formation of nanocrystals with high degrees of synthetic control	[8]
Biomimetic synthesis	It utilizes bio-inspired molecules to direct the synthesis of nanocrystals which can produce complex structures under mild conditions	[9]
Small adsorbate-assisted shape control	It focuses on the use of small molecules as the capping agent to fabricate nanostructures with well-defined surface	[10]

Table 1.2 Summaries of recent reviews focusing on various kinds of nanomaterials

Nanomaterials	Samples	References
Metals	Au, Ag, Pt, Pd, PtRu, AgPt, PtPd, Au/Pd, Ag/Au, PtPb, Pt–M (M = Co, Fe, Ni, Pd), M–Zn (M = Au, Cu, Pd), etc	[14]
Metal oxides	γ-Fe_2O_3, Fe_3O_4, ZnO, CoO, MnO, ZrO_2, TiO_2, etc	[15]
Metal chalcogenides	ZnS, Bi_2S_3, Cu_2S, Fe_7S_8, Fe_3S_4, NiS_2, β-In_2S_3, Ag_2Se, $CoSe_2$, β-FeSex, SnSe, Sb_2Se_3, GaSe, Bi_2Te_3, GeTe, etc	[16]
Lanthanide-doped nanocrystals	$NaYF_4$:Ln_3+, LaF_3, SmF_3, YF_3, CaF_2: Yb_3+/Er_3+, BaF_2: Eu_3+, $LaPO_4$, $LnVO_4$ (Ln = Y and lanthanides), etc	[17]

Table 1.3 Summaries of recent reviews focusing on various kinds of nanostructures

Nanostructures	Structural features	References
1D nanomaterials	1D nanomaterials including wires, tubes, belts, and rods have lateral dimensions between 1 and 100 nm	[18]
2D nanosheets	2D nanomaterials show unusual properties due to their ultra-thin thickness and 2D morphology	[19]
Heteronanocrystals	Heteronanocrystals are composed of two or more materials which have close interaction	[20]
Core/shell nanocrystals	Core/shell nanocrystals comprise a core (inner material) and a shell (outer layer material)	[21]

on various kinds of nanostructures including one-dimensional (1D) nanomaterials [18], 2D nanosheets [19], heteronanocrystals [20], and core/shell nanocrystals [24], which provide a comprehensive understanding of controllable synthesis of nanocrystals for readers.

During the synthetic process of nanocrystal, the nuclei will be generated when the cluster gets to a critical point and then develops into seeds and nanocrystals by the addition of generated atoms. The growth process can be experimentally monitored, and consequently has been well understood and controlled. Comparatively, it is difficult to catch the crystals at birth, due to the fast rate of the nucleation process and the extremely small grain size of the nuclei [25]. Despite these problems, many efforts have been devoted to investigating the nucleation process. As early as 1950, LaMer and co-workers first proposed a nucleation mechanism for the liquid-phase synthesis of nanoparticles [26]. Soon after, Myerson et al. tried to deeply understand the nucleation process of nanocrystals through theoretical calculation and further developed the nucleation theory of nanocrystals [27, 28]. However, it has always been a great challenge to actually monitor such a small object and the rapid process in real space and time. Fortunately, the rapid development of in situ characterization techniques has greatly promoted the further understanding on the nucleation process recently. Electron microscopy is a valuable and effective tool which helps us to synchronously observe the chemical composition, size, bonding, and electronic structure of nanocrystals. However, utilizing the conventional EM technique to study the nucleation processes of nanocrystals still has some disadvantages. On one hand, the poor spatial resolution of conventional EM is incapable of distinguishing nanoclusters whose sizes are below 1 nm. On the other hand, conventional TEM is usually an ex situ technique, so it cannot provide real information about spatial resolution and temporal resolution. Due to the development of aberration correction technologies and in situ characterization techniques, in situ electron microscopy has emerged [29]. It is a powerful tool to synchronously provide the real information about spatial resolution and temporal resolution [30]. Because the nucleation of nanocrystals occurs in solution, how to observe the dissolved samples in situ is also an issue needs to be considered [31].

When a series of technical problems have been solved, Alivisatos's group first utilized the in situ TEM technique to monitor the nucleation and growth process of platinum nanocrystals [32]. They observe minutely the different steps of nanocrystal growth, which is the first demonstration on nanocrystals nucleation and growth process in liquid. They found that the way particles accumulate can directly affect the structure of the resulting Pt particles in the nucleation process (Fig. 1.2a, b). It is the first time to directly observe the nucleation and growth of nanocrystals in the lab. Shortly afterward, this group has extended this in situ TEM method to atomic-level resolution imaging by introducing a new type of liquid cell based on the entrapment of a liquid film between layers of graphene (Fig. 1.2c–e). It allows us to have a clearer understanding on the nucleation and growth process [33]. Other external stimuli such as illumination-, heat-, and magnetic fields-induced structural evolution could also be fully studied by in situ technique. Direct observation from the precursors to nanocrystals could greatly enhance our comprehension of

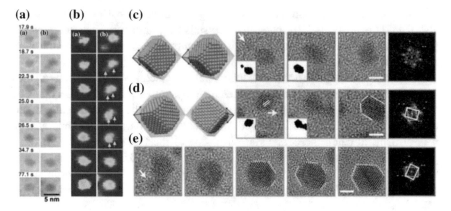

Fig. 1.2 Observation of the nucleation and growth process of Pt nanocrystals by in situ electron microscopy. **a, b** From [32]. Reprinted with permission from AAAS. **c, d, e** From [33]. Reprinted with permission from AAAS

nucleation mechanism and the following growth process. Furthermore, we can further design more complex catalysts under visible window by controlling the nucleation and growth of nanocrystals.

In the stage of generation of initial nuclei of metal cluster, it is unclear if the precursor compound is fully reduced into zero-valent atoms, or just partly reduced to metallic species which tend to form small clusters or nuclei. So, the characterization of cluster at atomic level is very important for the in-depth understanding of the nucleation process. Thereby, the in situ spectroscopic techniques are worthy of our attention, such as in situ X-ray absorption fine structure (XAFS) spectroscopy [34]. XAFS spectroscopy is a technique sensitive to the spatial information, and the resolution of this technique is atomic level. This technique provides the most authentic spectral signal of the structural transformation in the nanocrystalline nucleation process, and these spectral signals are closely related with the geometric and electronic structures of the objects observed [35].

As shown in Fig. 1.3, Wei's group successfully developed an in situ time-resolved XAFS technique setup with continuous-flow mode, observed the nucleation and growth process of Au nanoparticles, and monitored the structural transformation over time in situ from the precursor to Au clusters under 1 nm and then to the grown-up Au NCs. The entire reduction process of Au nanoparticles is completely recorded by utilizing this in situ technique. By combining the information given by in situ spectroscopy and reasonable assumption, the nucleation mechanism of Au can be probably inferred was extended by combining [36]. Furthermore, this newly developed in situ XAFS measurement was extended by combining UV-vis spectroscopies [37]. And this combination could permit a more detailed and complex comprehension on structural information and species evolution in a nucleation process. These in situ techniques can also be used to elucidate the nucleation pathways during the Pt NCs growth besides Au. For example, the different nucleation styles caused by different reducing conditions eventually

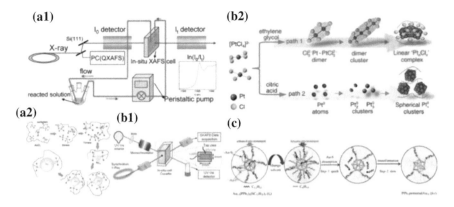

Fig. 1.3 a1 Schematic presentation of the experimental setup employed for in situ QXAFS measurements. **a2** A schematic representation of the formation process of Au NCs. Reprinted with the permission from Ref. [36]. Copyright 2010 American Chemical Society **b1** Experimental setup of the multi-in situ techniques combining UV-vis and quick-XAFS. **b2** Schematic representations of reaction pathways manipulated by reductants. Reprinted with the permission from Ref. [37]. Copyright 2012 American Chemical Society. **c** Schematic illustration of the structural transformation of Au NCs in different solvent. Reprinted with the permission from Ref. [38]. Copyright 2012 American Chemical Society

lead to different morphologies obtained in the nucleation process of Pt. The formation of one-dimensional "Pt_nCl_x" will lead to the shape of nanowires and the initial spherical "Pt_n^0" will result in the formation of nanospheres. Not only the transformation information of the geometric structures but also the transformation information of the electronic configurations of metal nanoclusters can be well addressed by this technique [38]. This technique provides a temporal resolution, and the information signal of geometric structure and electronic structure in atomic resolution. Thereby, we firmly believe that the development of in situ X-ray absorption fine structure spectroscopy provides a powerful means to research the nanocrystalline nucleation process in the future. Only by fully understanding the nanocrystalline nucleation process, we are able to better design nanocrystalline catalysts we need with specific geometric and electronic structures.

Because bimetallic synthesis involves the nucleation and growth process of two metals, we need to premeditate a lot of factors as follows:

Reduction potential: Standard reduction potential (SRP) can be utilized to measure the likelihood that a metal precursor will be reduced in aqueous solution. The higher SRP, the easier the metal ions can be reduced. Generally speaking, the SRP gives a trend of reducibility of different metal precursors. The complexation of the ligands and the metal ions can lower the concentration of metal ions and make the reduction of the metal cations more difficult. For example, halide ions in the synthesis of metallic nanocrystalline are common species which can reduce the redox potential of the metal. So, we can easily conclude that using ligands that strongly coordinate with the metal ions in the synthetic process can significantly increase the difficulty of metallic nucleation and decrease the reduction rate of the metal cations. These ligands often strongly interact with a certain set of crystal facets. So, the growth rate on the facets capped by ligands will be greatly slowed down, and the final metallic nanostructures will expose these stabilized facets [39]. The SRPs of noble metals are usually much higher than those of non-noble metals. If we want to synthesize alloy NCs containing noble metals and non-noble metals, we need to use ligands to coordinate strongly with the noble metal ions to lower their reduction potential. Recently, our group has reported that as-reduced noble metals preferentially transfer part of electrons to non-noble metals in the solution system, which could facilitate the reduction of non-noble metals. Under different conditions, we can also obtain bimetallic hybrid or core-shell structures [40]. In the reduction reaction, the presence of O_2 in the solution is another factor that should be taken into consideration. In the presence of oxygen, as-reduced metal atoms can be oxidized back to metal ions, as is called the oxidative etching process. Actually, ligands that strongly coordinate with the metal cations can also facilitate the oxidative etching process. Because the transformation from metal-salt to metal element is essentially a reducing process, so the reduction potential is a necessary factor we need to consider in the synthesis of bimetallic materials. Furthermore, reduction potential can affect the replacement reaction between two metals. For example, the replacement reaction can easily occur between the metal ions with high electronegativity and the elemental metals with low electronegativity. So during a reduction reaction, the galvanic replacement process and the oxidative etching process probably exist simultaneously.

Interfacial energy: If we try to synthesize bimetallic core-shell structure, we must consider the interfacial energy between two metals. If the interfacial energy is too high, the shell metal tends to adopt the isolated island growth model. In contrast, the shell metal tends to adopt the layered growth model outside the core metal when the interfacial energy is relatively low. As shown in Fig. 1.4, the left model stands for island growth and the right model stands for layered growth [41]. If we introduce these two growth models into the synthesis of bimetallic nanocrystals, we should consider the lattice match between two sets of metal lattices correspondingly.

Au@Ag and Pt@Pd core-shell NCs were prepared in some early works of Yang and co-workers. The lattice mismatch of Au–Ag and Pt–Pd is very small, i.e., 0.25 and 0.77 %, respectively [42]. Tian and co-workers first realized the preparation of Au@Pd core-shell NCs despite the lattice mismatch of 4.88 % [43]. Though the

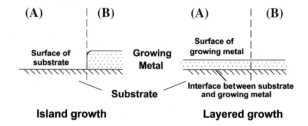

Fig. 1.4 Schemes of island growth model and layered growth model. Reproduced from Ref. [41] by permission of The Royal Society of Chemistry

lattice mismatch of Au–Pt (4.08 %) is smaller than that of Au–Pd, Au@Pt core-shell NCs with intact single-crystalline shells still cannot be synthesized. So we should consider not only lattice mismatch between two sets of lattices (should be smaller than 5 %) but also the bond between overlayer and substrate should be stronger than the bond in the overlayer. In a very recent work of Xia and co-workers, Cu could be epitaxially deposited on Pd seeds to form Pd@Cu core-shell nanostructures, despite the lattice mismatch between Cu and Pd is 7.1 % [44]. This result may originate from the strong affinity between Pd and Cu. Another driving force is the usage of hexadecylamine at the surface of Cu can decrease the surface free energy of Cu, which facilitates epitaxial growth. If the growth of the shell metal in core-shell NCs follows the layered growth model, the shell metal will adopt the same crystal growth direction and manner of atomic arrangement at the interface as the inner metal. And if the growth of the shell metal follows the island growth model, core-shell NCs can still be obtained under strongly reducing conditions, but usually with polycrystalline nature. As a result, lattice mismatch will generate a tension so as to facilitate the generation of defects and twins.

Reduction rate: The reduction rate of metal precursors is affected by many factors, including the redox potential of precursors, the reducing capacity of a reductant, and reaction temperatures. The reduction rate of metal precursors can be modulated by using different reductants and changing its concentration. When the reduction rate is slow, the nucleation and growth processes will be under near-equilibrium conditions. Otherwise, the fast reduction rate will break down near-equilibrium conditions. The metal precursor with high redox potential will be reduced preferentially rather than the other precursor under near-equilibrium conditions. This will lead to the different growth order of two metals, and the bimetallic NCs tend to form core-shell structure. In contrast, two metals prefer to form an alloy structure under equilibrium conditions. Similarly, the nucleation rate also plays an important role in modulating the surface structure of the metal. In the nucleation stage, a slow reduction rate favors the formation of multi-twinned seeds. Once twinned seeds appear, the energy barrier for the reconstruction from a twinned crystal to a single crystal is very high. If the reduction rate is further increased, it will lead to explosive nucleation, and a large amount of tiny crystallites will be generated simultaneously. These twins or defects own high surface energy and can further grow through collisions between nuclei. Therefore, the NCs obtained in this situation are thermodynamically unfavored.

Facet-specific capping agents: To synthesize bimetallic NCs with well-defined shapes, facet-specific capping agents are indispensable. These agents can be the counter ions in the precursors, solvents, salts, surfactants, polymers, gas molecules, and so on. In aqueous-phase synthesis, halide ions are often used. For example, Br− and I− have a significant effect as the {100} facet-specific capping agents for most of the face-centered cubic-structured metals [45]. In the synthesis of NCs, several steps involve the facet-specific capping agents. On one hand, capping agents can lower the surface energy of NCs and stabilize the specific facets. On the other hand, capping agents also can decrease the growth rate on certain facets, due to the selective binding effect on these facets. In other words, the

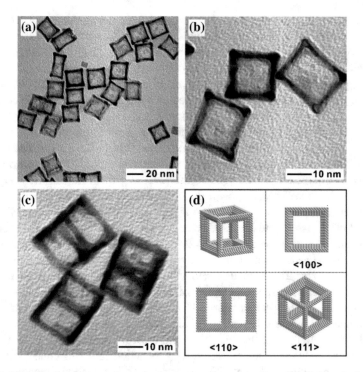

Fig. 1.5 The Rh cubic framework obtained using capping agent. Reproduced from Ref. [46] by permission of John Wiley & Sons Ltd

facet without capping agent will vanish in the growth process because of the faster growth rate, while the facet with the protection of capping agent will be retained in the end. The former effect is discussed in a thermodynamic control regime and the latter effect is related to kinetic control. Recently, it has been found that the facet-specific capping agent can control the oxidative etching process. For example, a capping agent we utilized can prevent the specific facet from contacting with O_2, and this leads to the occurrence of the oxidative etching on other unprotected facets. In a recent work of Xia and co-workers, a novel Rh cubic frame structure can be obtained by this synthetic strategy [46] (Fig. 1.5).

1.1.2 Nanocrystal Catalysis

For a half century, fruitful achievements have been made in the field of nanocatalysis by scientists from nanoscience, catalysis, organic chemistry, and other areas. Although the use of homogeneous catalysts still play a dominating role in organic synthesis, heterogeneous nanocrystals exhibit great potential in developing green and low-cost catalysts due to the ease of separation and the possibility

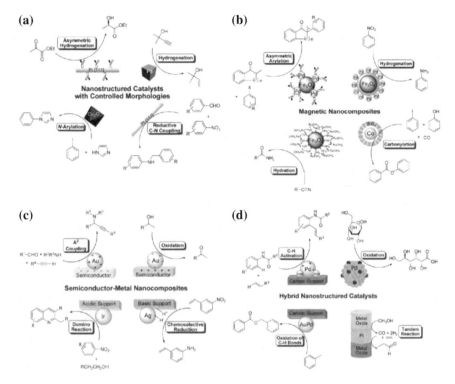

Fig. 1.6 The schematic diagram of different nanostructures in nanoparticle catalysis. Reproduced from Ref. [47] by permission of Royal Society of Chemistry

for continuous processing and catalyst recycling. The nanocrystal catalysis often undergoes a process that the reactant molecules adsorb on the surface of catalysts. Therefore, nanocrystal catalysis is considered to be surface and interface sensitive. Systematic study involving the control of the surface and interface plays an important role in elucidating the structure–property relationship in catalytic process. Considering the dominating role of surface atom arrangement in the catalytic performance, it is of great significance to develop synthetic strategies to sophisticatedly design and tune the surface and interfacial structures of nanocrystal catalyst. Based on the rapid development of nanoscience, we have gained great progress in the rational design of nanocatalysts with controlled size, composition, and structure by tuning support materials, capping agents, and so on.

In order to achieve more excellent performance of nanocatalysis, various structures of nanocatalysts have been synthesized and divided into four categories [47]. As shown in Fig. 1.6, (1) nanocrystal catalysts with well-defined morphologies and surface facets [48–50]: Catalytic properties of nanocatalysts can be improved due to the existence of unsaturated atoms in particular crystal facets. If this surface is modified with chiral ligands, we may achieve stereoselectivity in the catalytic reaction. (2) Magnetic separable nanocomposites [51–53]: These catalysts can participate in the catalytic cycle or support the active ingredient as a support. These nanomaterials tend to have good magnetic properties and thus can be separated for recycling. (3) Semiconductor-metal nanocomposites [54–56]: There exists electron transfer in such structure between the semiconductor and metal. As such, the different catalytic behavior can be controlled by modulating the electronic structure of the metal surface. The metal oxides are very common in this type of material, exhibiting significant synergistic effect with metal and playing an important role as a support. (4) Hybrid nanostructured catalysts [57, 58]: The scope of this structure is relatively widespread. Those structures such as composite structures of metal and oxidant (POMs), heterostructures of different metals, and bimetallic or polymetallic structures can be assigned to this category.

As the size of nanocrystal is reduced to the sub-nanometer regime, the nanocluster will exhibit a relative different catalytic behavior. As a classical example, Haruta et al. reported that the supported Au nanoparticles with their sizes being reduced to sub-10 nm exhibited a distinct activity, especially for CO oxidation [59] (Fig. 1.7). The chemical properties of metals, even the well-known inert metal Au, would drastically change when their sizes reduce to a certain range. So, we call it as size effect. In 2005, our group discovered that the CeO_2 nanocrystals with different crystal surface exhibited different activities for CO oxidation [60]. Such a series of important findings have inspired people to investigate the real structures of catalysts in nanoscale and reconsider how the active sites activate the substrates and accelerate reactions. Nowadays, people are still thinking about how size, composition, and structure affect the catalytic behavior [61] (Fig. 1.7). Thereby, how to control the active sites and catalytic dynamics has drawn more and more attentions. Implanting nanoscience to catalysis could advance the synthesis of catalysts, deepen the understanding of the catalytic processes, and open a new area for the catalysis research.

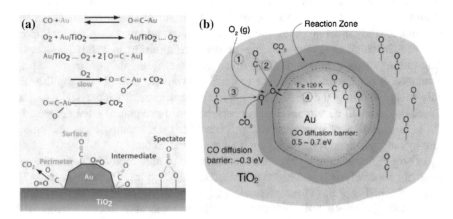

Fig. 1.7 The schematic diagram of Au/TiO$_2$ catalyzed CO oxidation; the left: Haruta model. Reprinted from Ref. [59], with kind permission from Springer Science Business Media the right: Yates model. From [61]. Reprinted with permission from Elsevier Ltd

The finding on size effect in Au catalysis mentioned above opens up the development of nanocatalysis. Under multidisciplinary cooperation, the development of nanocrystal catalysts has gained considerable progress. A variety of nanomaterials have been synthesized and successfully applied in the fields of photocatalysis [62, 63], fuel cells [64–66], lithium-ion battery [67, 68], heterogeneous catalysis [69], and organic catalysis [70, 71]. Even for a simple CO oxidation reaction, there are a lot of questions of about CO adsorption and O_2 activation. However, to date, there are still arguments on the essence of the reaction process and mechanism even for the most widely studied catalytic system. Haruta and Yates et al. reported that electron transfer exists at the interface between Au and TiO_2, respectively [72, 73]. The oxygen defects at the interface play a key role in O_2 activation. They put forward their own understanding of the reaction process by utilizing the model reaction. However, several fundamental questions of Au catalysis, for example, the authentic active sites of Au nanoparticle in a series of reactions and the relationship between microstructure and macroscopic catalytic properties are still unsolved. In fact, researchers working in the fields of catalysis and materials science have tried to explain these problems, but they hold that the quantitative structure–activity relationship of nanocrystal catalysts is interrelated with many factors, including: (1) the defects of the support surface, (2) positive metal atoms or clusters on the carrier surface, (3) unsaturated-coordinated metal atoms, (4) neutral metal atoms on the carrier surface, (5) the lattice strain in the metallic NPs, and (6) metal–nonmetal transitions of metal NPs [74–77]. These factors are strongly related to each other. Take Au as an example, at the atomic level, the physico-chemical property of Au NPs could be described with the following three parameters: the Au–Au interatomic distance, the coordination property of Au atoms, and the binding energy of Au atoms. The first two parameters describe the geometric structure of Au, and the third describes the electronic structure. As for nanocrystal catalysts, it is believed that the special properties and activities are strictly related to the geometric and electronic structures.

1.1.3 Pt–Ni Bimetallic Catalysts

As a novel heterogeneous catalyst, the bimetallic catalyst has attracted more and more attention due to its unique electronic and geometric properties. Since the study of C–H compound reforming reactions in 1960 [78], the bimetallic catalyst has been widely applied in the organic catalytic reactions [79], electrocatalytic reactions, and other fields [80]. In order to further explore the intrinsic mechanism for the excellent properties, the surface structure of bimetallic catalyst has been extensively studied. It has been generally concluded that the surface electrons and the geometric structure of bimetallic materials are completely different from the corresponding monometallic materials [81, 82]. The second metal introduced into the bimetallic catalyst can modify the surface or subsurface structures and accordingly affect the surface properties significantly. Because the study on nanocrystal catalyst has just emerged, the research on bimetallic catalyst is still challenging and needs further attention. It is found that two factors will affect the geometric and electronic structure of bimetallic catalysts. First, the introduction of heteroatoms and heteroatomic bond can considerably affect the electronic structure of metals. Second, the geometric structure of bimetals is completely different from that of monometallic metal. For example, the geometric structure including the average bond length variation and surface tension change also can affect the electronic structure through the orbital overlaps.

Many types of bimetallic materials have showed excellent catalytic activity and selectivity in catalysis field. For example, Au–Ag bimetallic catalyst shows sound synergistic effect and better catalytic activity than any monometal in catalytic CO oxidation reaction [83]. In this example, CO is adsorbed on Au, and the activation of O_2 prefers to occur on nearby Ag atoms. Some encouraging progress has recently been achieved in the study on non-noble metal bimetallic catalysts. For example, Ni–Cu bimetallic catalyst exhibited outstanding performance in $NaBH_4$ hydrolysis reaction to generate H_2 [84]. Bimetallic Pt/Ni material particularly attracted our attention due to the following reasons. (1) Its surface structure has been extensively studied than other bimetallic catalysts, a lot of researches including DFT theoretical calculation and material synthesis have been widely conducted. (2) It has been the most widely applied material in energy-related fields, such as organic catalytic reactions and electrocatalytic reactions.

In the past few decades, lots of methods have been developed to study the surface structure of bimetallic materials, such as DFT calculation, ultra-high vacuum (UHV) experiment, and liquid-phase synthesis. Face-centered cubic metals with different crystal facets exposed typically have different atomic arrangements on their surface layers (Fig. 1.8). Different arrangements will lead to different electronic and geometric properties of metal surfaces [41]. For example, the Pt/Ni bimetallic (111) facet shows three kinds of structures, Ni atoms can influence the electronic and geometric structure of Pt in different ways: Ni atoms occupy the outermost layer, and Pt atoms lie in the innermost part; some Ni atoms diffuse into the second outer layer; Ni atoms completely diffuse into the second outer layer, and the surface layer is crowded by Pt (Fig. 1.9). DFT calculation results have clearly shown that these three kinds of structures can significantly affect the d-band center of the bimetallic surface, thereby regulating the adsorption behavior of the reactant molecules on bimetallic catalyst surface and affecting its catalytic properties [85].

A heterogeneous catalytic process is usually divided into three phases: adsorption of reactant molecules on the surface of catalyst; reaction on the surface; desorption of products from the catalyst's surface. Numerous studies have indicated that the catalytic properties of Pt–Ni bimetallic catalyst are closely related to its surface structures.

Proper adsorption capacity can optimize the catalytic activity of catalysts. In the case of poor absorption, the reactant molecules can hardly be adsorbed on the catalyst surface and further activated, resulting in the failure of reaction. Moreover, selectivity of catalytic reaction is also related to selective adsorption

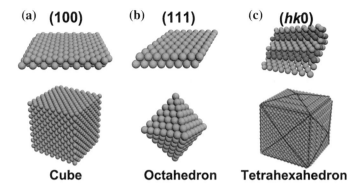

Fig. 1.8 Schematic of some examples of atomic arrangement on the surface of face-centered cubic (fcc) metal crystals with different shapes. Reproduced from Ref. [41] by permission of The Royal Society of Chemistry

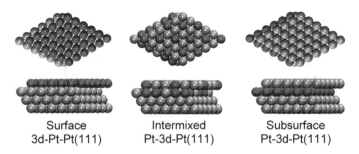

Fig. 1.9 Schematic of facets of different structure Pt/Ni bimetallic (111) facet. Reprinted with permission from Ref. [85]. Copyright 2012 American Chemical Society

of reactants and intermediates. The adsorption capacity is related to two factors. One is the arrangement manner of atoms on surface. The adsorbed molecules will adjust their configuration to match the surface atomic arrangement. So the adsorption energy will become weak and surface adsorption will take place. The other factor is the electronic structure of catalyst, especially the d-band center, which matters a lot. There is a linear relation between reactant molecules' adsorption capacity and the location of d-band center of catalyst. These two aspects are interrelated and mutually influenced. Further, these two aspects can be both influenced by the atomic arrangement of the bimetallic catalyst surface. So the regulation of catalysts' surface and controlled morphology synthesis can help to obtain bimetallic catalyst with high activity and selectivity. According to the study reported by Goodman group, Pd deposited on the surface of Au (100) facet and (111) facet shows different reaction activities for the synthesis of vinyl acetate. The lattice of Pd deposited on (100) facet is mainly 4.08Å and it matches the adsorption of vinyl and acetate on the surface better than 4.99Å shown on (111) facet [90]. It is critical to find a proper way to achieve the controlled synthesis of Pt–Ni alloy

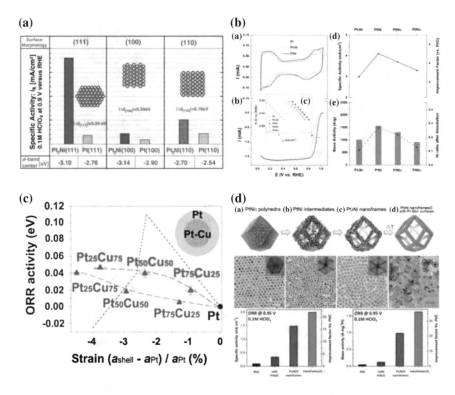

Fig. 1.10 a The comparison of ORR catalytic activity of different crystal facets of Pt–Ni alloy. From [86]. Reprinted with permission from AAAS; **b** The relationship of ORR reactivity related with the different components of the Pt–Ni alloy. Reproduced from Ref. [87] by permission of John Wiley & Sons Ltd; **c** The relationship of ORR catalytic of core-shell structure related with surface tension. Reprinted by permission from Macmillan Publishers Ltd: Ref. [88]. Copyright 2010; **d** The excellent reaction catalytic activity of ORR of Pt–Ni frame structure. From [89]. Reprinted with permission from AAAS

and optimize its *d*-band center for regulation of its performance in catalytic reaction. It is the cathode oxygen reduction reaction that hinders the fuel cell development [85]. In this reaction, the competitive adsorption of OH⁻, O₂, and other species on the catalyst determines the rate of the reaction. As was demonstrated by the comparative analysis of ORR activity on different crystal surfaces of Pt₃Ni conducted by Stamenkovic et al. in 2007, the bimetallic surface with a segregated Pt surface was much better than pure Pt surface. As for the different crystal facets, different surface structures result in different *d*-band centers. The surface adsorption energy of OH⁻, O₂, and other species can be adjusted. The activity of different surfaces can be ranked as follows: Pt₃Ni {100} < Pt₃Ni {110} < Pt₃Ni {111} [86] (Fig. 1.10a). According to their findings in 2010, different components of Pt–Ni alloy can influence the atomic arrangement on surfaces and thereby regulate its activity in the oxygen reduction reaction [87] (Fig. 1.10b). As was revealed in the study conducted by Peter Strasser et al. in 2012, the way of alloy synthesis can help to adjust the surface tension of bimetallic surface [88](Fig.1.10c). Once the surface lattice is extruded and contracted, the bimetallic material's d-band center will be downshifted. So the change of surface tension can also affect the bimetallic material's d-band center and catalytic performance. The shell with Pt-segregated surface can lead to the downward shift of the d-band center and reduce oxidative species' adsorption on the surface of catalysts, ultimately affecting the catalytic activity of oxygen reduction reaction. According to Chen et al.'s [89] work published on Science, the structure of Pt_3Ni can increase the specific surface area and provide a 3D pore structure. The bimetallic surface realized by a dealloying method is also Pt-segregated. Such a frame-structured bimetallic catalyst embracing 3D pore structure and segregated Pt surface has shown the best activity and stability in catalytic oxygen reduction reaction so far (Fig. 1.10d).

In nanocrystal catalysis, the selectivity of reaction should also be considered. For example, in electrocatalytic oxidation of methanol, methanol oxidation is generally realized in two manners. One is the direct oxidation of methanol to CO_2 without the formation of CO species. In the other way, CO intermediate species are generated first, and then further oxidized into CO_2. For the Pt-based bimetallic catalyst, the direct path of methanol oxidation on the surface of {111} is thermodynamically favorable, while the indirect path on {111} surface requires more energy for activation. As for the {100} facet, both the direct and the indirect paths are likely to occur, because of the similar activation energy. So, CO and other species will exhibit more obvious poisoning effect on the {100} crystal surface, and the stability of Pt-based bimetallic materials with {111} surface is much higher than that of the catalyst with {100} surface exposed [91] (Fig. 1.11). Not only the crystal face, but also the microstructure of catalyst surface, including twin crystal, defects, and high index surface, can increase the lattice tension on the catalyst's surface and thereby change the adsorption energy of reactants and affect the selectivity of catalysts [92–94].

As implied by the above results, how to control the surface structure of Pt–Ni bimetallic catalyst at nanoscale via synthetic methodology and accordingly adjust its special geometry and electronic structure has become critical for the research of Pt–Ni catalyst. In recent decades, in order to obtain the bimetallic structure with controllable size, composition, and structure, lots of universal and effective methods have been developed, including coreduction, thermal decomposition, seed growth, replacement reaction, and precious metal induction of non-noble metal reduction [95] (Fig. 1.12). Moreover, the development of synthetic technology also provides various methods for the synthesis of functionalized Pt–Ni catalyst, and various Pt–Ni bimetallic nanomaterials with controllable structure.

In Fang group's study, using oleylamine and oleic acid as the surfactants, CO as crystal face inhibitor, Pt–Ni bimetallic octahedron exposing {111} facet and cubes exposing {100} facet can be obtained, respectively (Fig. 1.13). The catalytic performance of Pt–Ni alloy with different structures toward oxygen reduction reaction was systematically investigated [96, 97].

In addition, Yang's group has also reported the synthesis of icosahedral Pt–Ni alloy exposing {111} facet and truncated octahedral Pt–Ni bimetallic alloy exposing {111} and {100} facets [98, 99] (Fig. 1.14).

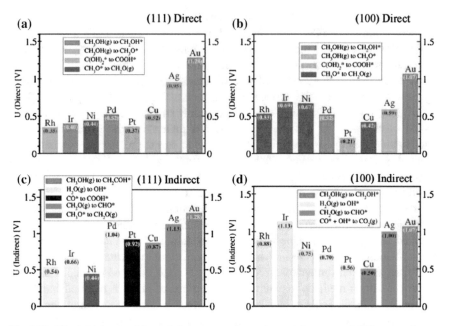

Fig. 1.11 The initial potentials and the energy of rate-determining step of different paths on metal surfaces in methanol oxidation reaction. Reprinted with the permission from Ref. [91]. Copyright 2009 American Chemical Society

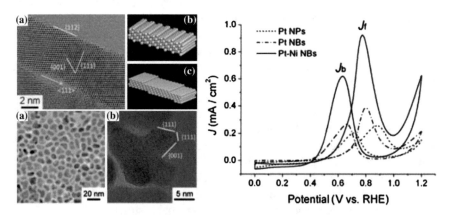

Fig. 1.12 The catalytic performance of Pt–Ni alloy with surface defects for methanol oxidation. Reproduced from Ref. [95] by permission of The Royal Society of Chemistry

We found that the above-described Pt–Ni bimetallic catalysts are mainly synthesized in nonpolar solvent. Due to the strong binding effect of oleylamine and oleic acid surfactants, it is not easy to remove them from the surface of NPs. These long-chain ligands on metal surface can to some extent reduce the catalytic activity of the catalyst. According to Strasser's research, by adding the support into the

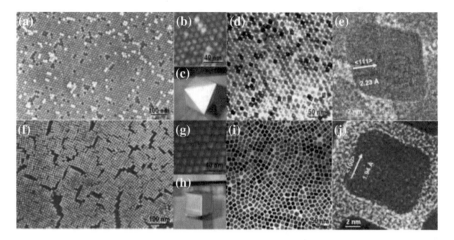

Fig. 1.13 The synthesis of Pt–Ni octahedron and cube. Reprinted with the permission from Ref. [97]. Copyright 2010 American Chemical Society

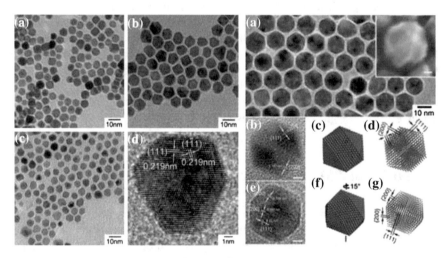

Fig. 1.14 The synthesis of Pt–Ni truncated octahedron and icosahedron. The *left* Reprinted with the permission from Ref. [98]. Copyright 2010 American Chemical Society. The *right* Reprinted with the permission from Ref. [99]. Copyright 2012 American Chemical Society

synthetic system, they have successfully prepared a supported Pt–Ni octahedron catalyst in N,N-dimethylformamide without surfactants on its surface (Fig. 1.15). Moreover, they also reported that the as-prepared Pt–Ni bimetallic catalyst exhibits superior catalytic activity compared to catalyst whose surface capped with surfactant or ligand [100].

Our recent research has developed a dealloying method, with which we can convert Pt–Ni rhombic dodecahedron particles to porous Pt–Ni bimetallic catalysts

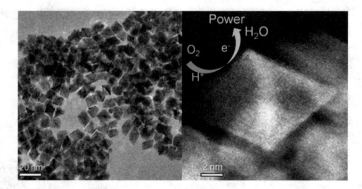

Fig. 1.15 Synthesis of surfactant-free Pt–Ni octahedron. Reprinted with the permission from Ref. [100]. Copyright 2010 American Chemical Society

utilizing the reaction of nitric acid with Ni (Fig. 1.16). We also found that the porous Pt–Ni bimetallic catalysts exhibit more superior electrocatalytic activity than the parent polyhedron because of the larger surface area [101].

And then, Erlebacher and Strasser have reported that Pt–Ni alloy can be dealloyed through a spontaneous activation process in acidic electrolyte solution. They have obtained porous Pt–Ni nanoparticles and concave Pt–Ni polyhedron with Pt-segregated surface, utilizing different precursors, respectively (Figs. 1.17 and 1.18). They also found that the catalytic activity of oxygen reduction reaction is greatly improved. They attributed this enhancement to larger surface area and controlled Pt surface segregation, respectively [102, 103].

1.2 Theme and Main Content of this Thesis

Nanocrystal catalysis is a young interdisciplinary field. From developing the synthetic methodologies to obtaining a functional nanocrystal catalyst, to evaluating its catalytic performance by practical catalytic reactions, and to explaining the relationship between the catalyst and its catalytic properties, they all require the deep understanding of the active sites of catalyst and its catalytic performance base on the electronic properties and geometric structures. In order to obtain practical catalyst with a novel structure, we tried to develop new synthetic methods and strategies to sophisticatedly modulate surface structure of the catalyst, and to further design and synthesize the nanocrystalline catalyst under the guidance of the nucleation and growth mechanism of nanocrystals. Using monodisperse nanocrystals with well-defined size, composition, structure as catalysts, we attempted to obtain the catalyst with more excellent catalytic properties and further understand the structure–activity relationship of the catalyst in the pivotal catalytic processes of organic catalysis and electrocatalysis.

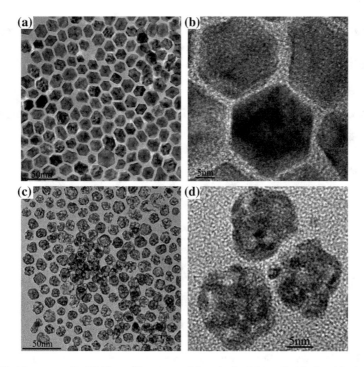

Fig. 1.16 The porous Pt–Ni bimetallic nanoparticles obtained from Pt–Ni rhombic dodecahedron by dealloying method. Reprinted by permission from Macmillan Publishers Ltd: Ref. [101]. Copyright 2011

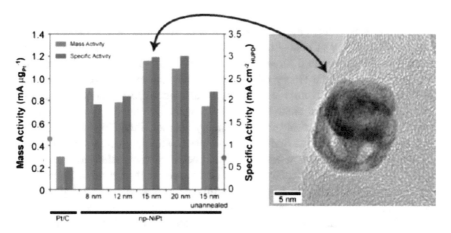

Fig. 1.17 Pt–Ni porous structure formed spontaneously in an acidic electrolyte solution. Reprinted with the permission from Ref. [102]. Copyright 2011 American Chemical Society

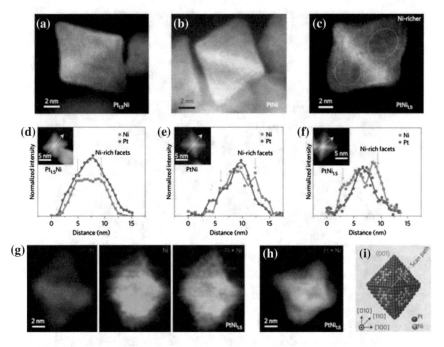

Fig. 1.18 Pt–Ni concave structure obtained in an acidic electrolyte solution. Reprinted by permission from Macmillan Publishers Ltd: Ref. [103]. Copyright 2013

1.2.1 Opportunities and Challenges of Nanocatalysis

Synthetic method: The past decades have witnessed the rapid development of nanoscience. Researchers have developed lots of synthetic methods and strategies to get a wide variety of nanostructures [104] (Fig. 1.19), which provide abundant material for us to screen a more efficient catalyst. Simultaneously, with the further development of the theory of nanocrystal nucleation and growth, and emergence of new characterization techniques, we also get a clear understanding of the structure of the obtained catalysts. However, with the dwindling of fossil fuels, there is an increasing demand for new energy sources, including wind, biomass, solar, and other new energy. Some energy-related catalytic processes, such as photolysis of water, biomass conversion and fuel cells, have attracted more and more attention. Development of catalysis science requires high-performance catalysts. Constrained by a series of factors including energy, environmental protection, and others, we have to develop catalytic reactions with low energy consumption, less pollution, and high added value, which require for catalysts with low cost, high activity, good selectivity, and long service life. Moreover, the development of catalysis science requires more sophisticated catalysts. Therefore, we have to continuously achieve new breakthroughs in developing new nanomaterials. As a simple example, for the metal catalyst, the traditional synthetic methods are not

Fig. 1.19 The schematic diagram of the nucleation and growth process of nanocrystals proposed by Lamer et al. Reproduced from Ref. [104] by permission of John Wiley & Sons Ltd.

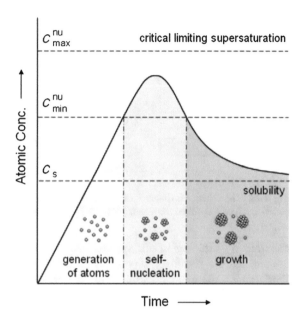

able to modulate the surface structure of nanocrystal catalysts at atomic level. Furthermore, there is a lack of appropriate synthetic strategies and theories for controlled synthesis of complex structures, especially those multilevel composite structures. Nowadays, we are able to utilize some techniques such as in situ spectroscopic and electron microscopy to investigate the nucleation process of nanocrystal. However, it is undoubted that our understanding of this complex and fast process is still very limited. Moreover, we have not had a mature theory to prompt better understanding of nanocrystal nucleation process. Furthermore, the manner of nucleation plays a significant role in affecting the crystal facet, defects, and growth pattern. So the deficient understanding restricts the ability of controlling the surface structure of nanocrystals precisely. Thereby, there are tremendous opportunities and challenges in the research of nanocrystal synthesis.

Catalytic mechanism: the research of the active sites of the catalyst has been a hot topic in the field of catalysis. In order to clearly explain the catalytic behavior of the active sites, it requires understanding on how their electronic and geometrical structures determine the activation of the molecule. However, to date, there are still many problems on the real catalytic mechanism of the active sites, even for the most widely studied supported catalysts such as traditional metal/metal oxide (Fig. 1.20). However, the understanding of the real active sites of metals is not enough. For example, where does the substrate adsorb, on the surface or interface? Where does the activation of molecules occur? Moreover, the understanding of the catalytic mechanism of homogeneous and heterogeneous nanocrystal catalyst and the real catalytic active sites is also not enough. What species is the real active site of catalyst, single atoms, clusters, or nanocrystals? Will the active species change in the reaction process? Although we are able to monitor the progress of some

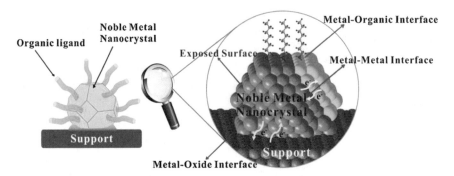

Fig. 1.20 The schematic diagram of the surface and interface structure of a conventional supported catalysts Reprinted from Ref. [105]. Copyright 2013, with permission from Elsevier

reactions by utilizing in situ spectroscopy, the information we obtained is very limited and the reactions which can be monitored are very simple and limited. Furthermore, the method for probing the nonperiodic amorphous structure is still limited to the solid-state NMR, EXAFS, and a few others. Verifying the structure of active site in situ is still a worldwide challenge. Fortunately, with the development of the understanding of nanocatalysis, we can generally predict that the sub-nano clusters are likely to be the real active sites. However, the exact type of these clusters and the process how they modulate among a series of transition states is unclear. Thus, the comprehensive description of an active site is significant, which is also related to the definition of nanocatalysis. Namely, how are these nano- and subnanoclusters different from the coordination compounds or bulk particles? What are the catalytic properties of a perfectly flat surface without any defects? Further advances in nanocatalysis call for clear answers to these questions.

Catalytic Properties: The concept of nanocatalysis has been introduced in various fields, such as photocatalysis, dye-sensitized solar cells, organic catalysis, biomass conversion, fuel cells and etc. Although nanocatalysts have been extensively studied in these fields, most of the studies still remain in the laboratory stage and the limited reaction efficiency restricts their practical applications. For examples, the efficiency of photocatalysis is very low. We cannot achieve nitrogen fixation under room temperature and atmospheric pressure as the plants. The conversion efficiency of solar energy is still very limited. The conversion efficiency of biomass energy is very low and the conditions are very harsh. These practical problems greatly facilitate the rapid development of nanocatalysis. Taking organic catalysis as a starting point, our goal is to summarize the current situation, opportunities, and challenges of nanocatalysis. In the field of organic catalysis, the application of nanocrystalline catalyst is also limited to some simple reactions, such as the coupling reaction and the oxidation/reduction reaction, and cannot be extended to complex organic reactions and the synthesis of the complex compounds [106]. In recent years, Toste and Somorjai et al. have innovatively utilized surface modification of nanocrystal catalyst to achieve the synthesis of a series of complex organic compounds. Previously, these reactions are not considered to be achievable by using heterogeneous nanocrystal catalyst. Thus, it is the first time to break the barriers between the heterogeneous catalysis and homogeneous catalysis [107]. In the field of the traditional organic catalysis, metal ions integrated with corresponding ligands are commonly used as catalyst. The active site of catalyst is traditionally considered to be single-center. The reactive characteristics of metal ion are modulated by tuning the interaction between it and the corresponding ligands. However, in recent years, a growing number of ground-breaking studies make researchers have different views on the argument that whether the active site in conventional homogeneous catalysis is the single-center metal or the multicenter clusters or nanoparticles. More recently, the remarkable catalytic performance of TON(10^7) and TOF (10^5 h^{-1}) for the ester-assisted hydration of alkynes has been achieved by Corma and coworkers in Spain [108] (Fig. 1.21a). And they consider that the Au cluster, formed in situ, is the real active site in this reaction. Additionally, the generation of Au clusters has been monitored by mass spectrometry and UV-visible spectra. A corresponding induction process has been observed by the catalytic kinetic curves. Above evidence proved that it is the Au clusters instead of Au ions playing the key role in this reaction. Because this system does not have other impurities, this result is very reliable. Thereby, we believe that the emergence of this work allows us to further understand the mechanism of the reaction. After this, a variety of similar works gradually appeared. Subsequently Shannon Stahl's group also reported that the Pd clusters, formed in situ, are the truly active substances in catalytic dehydrogenation of cyclohexanone [109] (Fig. 1.21b). All these findings continuously deepen our understanding of catalysis and also lay the foundation for the design of a catalyst which has industrial applications.

In summary, from the research on synthetic methodology of nanocrystalline catalysts, to the study on their catalytic mechanism applications, the field of nanocrystal catalysis is full of unknowns and challenges. To achieve a better blueprint of future research, scientists from different areas, e.g., materials science, organic chemistry, nanoscience etc., should work in close cooperation with each other. The success also relies on the arduous efforts of the younger generation.

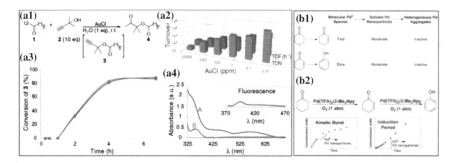

Fig. 1.21 (**a**) The hydration reaction of alkynes catalyzed by Au. From [108]. Reprinted with permission from AAAS; (**b**) the cyclohexanone dehydrogenation reaction catalyzed by Pd. Reprinted with the permission from Ref. [109]. Copyright 2013 American Chemical Society

1.2.2 The Main Contents of This Thesis

With the development of nanocatalysis, we will meet more and more challenges as well as opportunities. As shown in Fig. 1.22, we chose the Pt–Ni alloy as the main study object. We developed a new synthetic methodology to design and synthesize functionalized Pt–Ni bimetallic catalyst. Furthermore, we established a new catalytic system by precisely controlling the geometric and electronic structures, and further understood the structure–function relationship of the nanocatalyst.

Following is the general introduction to Chaps. 2, 3, 4, and 5:

In Chap. 2, we focus on the controllable synthesis and the structure–function relationship of the Pt–Ni alloy in a water-soluble reaction system. In recent years, the bimetallic alloys have attracted extensive attention in the fields of catalysis and energy. Abundant bimetallic nanocrystals have been prepared following the mature synthetic strategies in nonpolar solvent. But there were few reports about the bimetallic alloy synthesized in the polar solvent. However, there are enormous potential applications in biology and catalysis for the polar solvent system due to the favorable biocompatibility. In this work, we had synthesized a series of water-soluble alloys in the benzyl alcohol system, systematically controlling the shape of Pt, Pt_xNi_{1-x} $(0 < x < 1)$, and Ni nanocrystal with the utilization of some small molecules (aniline, benzoic acid, carbon monoxide). In the field of catalysis, it is always a problem to achieve the selective hydrogenation under mild conditions. In this chapter, we systematically studied the catalytic properties of the bimetallic nanocrystal toward the hydrogenation reactions of benzalacetone, styrene, and aniline. We found that the Pt–Ni alloy exhibited superior activity and selectivity relative to its monometallic counterparts, achieving the hydrogenation of the $C = C$ bonds in high yield at room temperature and atmosphere pressure. To further understand the composition–function relationship, we selected the catalyst $PtNi_2$ alloy with best performance, which achieved 100 % conversion with nearly 100 % selectivity for the hydrogenation of the C=C bonds. Furthermore, we developed a simple method to modify the surface of the catalyst, by which we could make the catalyst favorably dissolved in both nonpolar and polar solvents. In our method, Pt–Ni nanocrystals are capped with PVP which is an environmentally friendly

Fig. 1.22 The research ideas of this thesis

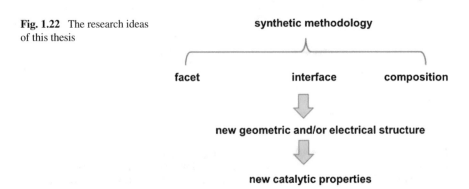

and nontoxic surfactant with favorable biocompatibility. We also demonstrated the addition of PVP would not affect the activity and selectivity of the catalysts. This finding lights up the path to further understand the catalytic mechanism and design efficient catalysts.

In Chap. 3, we tried to build the high-active surface of the Pt–Ni bimetallic catalysts. The atoms located at steps exhibited high reactivity during the catalytic process owing to their low-coordinated numbers, they could effectively decrease the activation energy of the reactions to accelerate the process. However, the problem that puzzles the researcher is how to effectively combine the surface defects with the bimetallic NPs. Herein, we have developed a controllable etching strategy to build surface defects on the surface of bimetallic NPs. It is generally accepted that dimethylglyoxime cannot react with Ni(0). But the presence of dimethylglyoxime can effectively facilitate the conversion of Ni(0) to Ni(II). We can also obtain a series of bimetallic nanocrystals with increasing degree of etching by controlling the percentage of the Ni in starting Pt–Ni alloy. Besides, we modulated a theoretical simulation for the etching process. Surprisingly, we observed the priority of the chemical etching process when the etching rate slowed down to a certain extent. Next, we evaluated the structure–activity relationship of the catalyst through several typical catalytic reactions. Remarkably, owing to the larger surface area and higher density of exposed atomic steps, the concave nanostructures exhibited superior activity over the octahedral Pt–Ni alloy particles for both the electrooxidation of methanol and hydrogenation of nitroarenes. This work also plays a crucial role in guiding us to design effective catalyst.

In Chap. 4, we invested efforts in studying the effects of defects on the surface of bimetallic nanocrystals, and tried to realize the controllable synthesis of the trimetallic catalysts via this effect. In view of the different rates of nucleation and growth between the different compositions in the polymetallic system, traditional synthetic methodologies such as coreduction and codeposition are becoming insufficient to completely control the synthesis of polymetallic nanocrystals. Interestingly, we have obtained a concave structure through selective etching of the more active metal Ni from the bimetallic Pt–Ni alloy. In the second procedure, concave seeds were redispersed in a Ni^{2+}-rich chemical environment, following which the shape recovery of nanocrystals can be observed. According to the experimental analysis and theoretical calculation, we profoundly make the assertion that high-density atomic steps and defects on the surface might facilitate the nucleation around it. Herein, utilizing the defects on the surfaces of the bimetallic nanocrystals, a predominant control of the composition and the shape of the third metal can be realized. Finally, we achieved the controllable synthesis of the trimetallic nanocrystals and systematically evaluate their structure–activity relationship toward some important reactions, expanding a new strategy to design trimetallic catalysts.

In Chap. 5, we devoted ourselves to establish the sophisticated trimetallic nanoframe catalyst. Metallic nanoparticles with a nanoframe structure, in which edge and corner atoms are joined together to create a skeletal frame, can serve as an ideal model catalyst. Considering the coordination environment around different geometrical sites, the corner and edge atoms exhibit higher activity than those at the flat surface, being liable to be attacked by other atoms. Herein, the third composition would be preferentially introduced into the corner and edge if we want to decorate the bimetallic nanoparticles with the third composition. We utilized chemical etching to obtain the nanoframe structure, while retaining the active sites (such as corner and edge). So, the hinge of the controllable synthesis of trimetallic nanoframe is to combine the galvanic and chemical etching together, which seems to be contradictory. Joyfully, we successfully obtained this novel trimetallic nanoframe structure by the successive strategy, combining the replacement reaction together with controllable etching. The subsequent catalytic model reactions demonstrated that this novel structure is an extremely ideal model catalyst, which achieves not only the optimization of the activity but also high selectivity and stability.

References

1. Eigler DM, Schweizer EK (1990) Positioning single atoms with a scanning tunnelling microscope. Nature 344(6266):524–526
2. Cushing BL, Kolesnichenko VL, O'Connor CJ (2004) Recent advances in the liquid-phase syntheses of inorganic nanoparticles. Chem Rev-Columbus 104(9):3893–3946
3. Wang X, Peng Q, Li Y (2007) Interface-mediated growth of monodispersed nanostructures. Acc Chem Res 40(8):635–643
4. Modeshia DR, Walton RI (2010) Solvothermal synthesis of perovskites and pyrochlores: crystallisation of functional oxides under mild conditions. Chem Soc Rev 39(11):4303–4325
5. Nadagouda MN, Speth TF, Varma RS (2011) Microwave-assisted green synthesis of silver nanostructures. Acc Chem Res 44(7):469–478
6. Sui R, Charpentier P (2012) Synthesis of metal oxide nanostructures by direct sol–gel chemistry in supercritical fluids. Chem Rev 112(6):3057–3082
7. Xu H, Zeiger BW, Suslick KS (2013) Sonochemical synthesis of nanomaterials. Chem Soc Rev 42(7):2555–2567
8. Yin Y, Talapin D (2013) The chemistry of functional nanomaterials. Chem Soc Rev 42(7):2484–2487
9. Chiu C-Y, Ruan L, Huang Y (2013) Biomolecular specificity controlled nanomaterial synthesis. Chem Soc Rev 42(7):2512–2527
10. Chen M, Wu B, Yang J, Zheng N (2012) Small adsorbate-assisted shape control of Pd and Pt nanocrystals. Adv Mater 24(7):862–879
11. Xu H, Zeiger BW, Suslick KS (2013) Sonochemical synthesis of nanomaterials. Chem Soc Rev 42(7):2555–2567
12. Liu Y, Goebl J, Yin Y (2013) Templated synthesis of nanostructured materials. Chem Soc Rev 42(7):2610–2653
13. Chiu C-Y, Ruan L, Huang Y (2013) Biomolecular specificity controlled nanomaterial synthesis. Chem Soc Rev 42(7):2512–2527
14. You H, Yang S, Ding B, Yang H (2013) Synthesis of colloidal metal and metal alloy nanoparticles for electrochemical energy applications. Chem Soc Rev 42(7):2880–2904
15. Kwon SG, Hyeon T (2008) Colloidal chemical synthesis and formation kinetics of uniformly sized nanocrystals of metals, oxides, and chalcogenides. Acc Chem Res 41(12):1696–1709

16. Gao M-R, Xu Y-F, Jiang J, Yu S-H (2013) Nanostructured metal chalcogenides: synthesis, modification, and applications in energy conversion and storage devices. Chem Soc Rev 42(7):2986–3017
17. Wang G, Peng Q, Li Y (2011) Lanthanide-doped nanocrystals: synthesis, optical-magnetic properties, and applications. Acc Chem Res 44(5):322–332
18. Yuan J, Xu Y, Müller AH (2011) One-dimensional magnetic inorganic–organic hybrid nanomaterials. Chem Soc Rev 40(2):640–655
19. Huang X, Zeng Z, Zhang H (2013) Metal dichalcogenide nanosheets: preparation, properties and applications. Chem Soc Rev 42(5):1934–1946
20. de Mello DC (2011) Synthesis and properties of colloidal heteronanocrystals. Chem Soc Rev 40(3):1512–1546
21. Ghosh Chaudhuri R, Paria S (2011) Core/shell nanoparticles: classes, properties, synthesis mechanisms, characterization, and applications. Chem Rev 112(4):2373–2433
22. You H, Yang S, Ding B, Yang H (2013) Synthesis of colloidal metal and metal alloy nanoparticles for electrochemical energy applications. Chem Soc Rev 42(7):2880–2904
23. Gao M-R, Xu Y-F, Jiang J, Yu S-H (2013) Nanostructured metal chalcogenides: synthesis, modification, and applications in energy conversion and storage devices. Chem Soc Rev 42(7):2986–3017
24. Chaudhuri RG, Paria S (2012) Core/shell nanoparticles: classes, properties, synthesis mechanisms, characterization, and applications. Chem Rev 112(4):2373–2433
25. Oxtoby DW (2000) Phase transitions: catching crystals at birth. Nature 406(6795):464–465 (a-z Index)
26. LaMer VK, Dinegar RH (1950) Theory, production and mechanism of formation of monodispersed hydrosols. J Am Chem Soc 72(11):4847–4854
27. Auer S, Frenkel D (2001) Prediction of absolute crystal-nucleation rate in hard-sphere colloids. Nature 409(6823):1020–1023
28. Erdemir D, Lee AY, Myerson AS (2009) Nucleation of crystals from solution: classical and two-step models. Acc Chem Res 42(5):621–629
29. Bordiga S, Groppo E, Agostini G, van Bokhoven JA, Lamberti C (2013) Reactivity of surface species in heterogeneous catalysts probed by in situ X-ray absorption techniques. Chem Rev 113(3):1736–1850
30. Klein K, Anderson I, De Jonge N (2011) Transmission electron microscopy with a liquid flow cell. J Microsc-Oxford 242(2):117–123
31. de Jonge N, Ross FM (2011) Electron microscopy of specimens in liquid. Nat Nanotechnol 6(11):695–704
32. Zheng H, Smith RK, Y-w Jun, Kisielowski C, Dahmen U, Alivisatos AP (2009) Observation of single colloidal platinum nanocrystal growth trajectories. Science 324(5932):1309–1312
33. Yuk JM, Park J, Ercius P, Kim K, Hellebusch DJ, Crommie MF, Lee JY, Zettl A, Alivisatos AP (2012) High-resolution EM of colloidal nanocrystal growth using graphene liquid cells. Science 336(6077):61–64
34. Colombi Ciacchi L, Pompe W, De Vita A (2001) Initial nucleation of platinum clusters after reduction of K_2PtCl_4 in aqueous solution: A first principles study. J Am Chem Soc 123(30):7371–7380
35. Wang Q, Hanson JC, Frenkel AI (2008) Solving the structure of reaction intermediates by time-resolved synchrotron x-ray absorption spectroscopy. J Chem Phys 129:234502
36. Yao T, Sun Z, Li Y, Pan Z, Wei H, Xie Y, Nomura M, Niwa Y, Yan W, Wu Z (2010) Insights into initial kinetic nucleation of gold nanocrystals. J Am Chem Soc 132(22):7696–7701
37. Yao T, Liu S, Sun Z, Li Y, He S, Cheng H, Xie Y, Liu Q, Jiang Y, Wu Z (2012) Probing nucleation pathways for morphological manipulation of platinum nanocrystals. J Am Chem Soc 134(22):9410–9416
38. Li Y, Cheng H, Yao T, Sun Z, Yan W, Jiang Y, Xie Y, Sun Y, Huang Y, Liu S (2012) Hexane-Driven icosahedral to cuboctahedral structure transformation of gold nanoclusters. J Am Chem Soc 134(43):17997–18003

39. Jana NR, Gearheart L, Murphy CJ (2001) Wet chemical synthesis of silver nanorods and nanowires of controllable aspect ratio electronic supplementary information (ESI) available: UV–VIS spectra of silver nanorods. Chem Commun (7):617–618

40. Wang D, Li Y (2010) One-pot protocol for Au-based hybrid magnetic nanostructures via a noble-metal-induced reduction process. J Am Soc 132(18):6280–6281

41. Gu J, Zhang YW, Tao FF (2012) Shape control of bimetallic nanocatalysts through well-designed colloidal chemistry approaches. Chem Soc Rev 41(24):8050–8065

42. Song JH, Kim F, Kim D, Yang P (2005) Crystal overgrowth on gold nanorods: tuning the shape, facet, aspect ratio, and composition of the nanorods. Chem-Eur J 11(3):910–916

43. Fan F-R, Liu D-Y, Wu Y-F, Duan S, Xie Z-X, Jiang Z-Y, Tian Z-Q (2008) Epitaxial growth of heterogeneous metal nanocrystals: from gold nano-octahedra to palladium and silver nanocubes. J Am Chem Soc 130(22):6949–6951

44. Jin M, Zhang H, Wang J, Zhong X, Lu N, Li Z, Xie Z, Kim MJ, Xia Y (2012) Copper can still be epitaxially deposited on palladium nanocrystals to generate core-shell nanocubes despite their large lattice mismatch. ACS Nano 6(3):2566–2573

45. Xia Y, Xiong Y, Lim B, Skrabalak SE (2009) Shape-controlled synthesis of metal nanocrystals: simple chemistry meets complex physics? Angew Chem Int Ed 48(1):60–103

46. Xie S, Lu N, Xie Z, Wang J, Kim MJ, Xia Y (2012) Synthesis of Pd–Rh core-frame concave nanocubes and their conversion to Rh cubic nanoframes by selective etching of the Pd cores. Angew Chem Int Ed Engl 51(41):10266–10270

47. Chng LL, Erathodiyil N, Ying JY (2013) Nanostructured catalysts for organic transformations. Acc Chem Res 46(8):1825–1837

48. Sawai K, Tatumi R, Nakahodo T, Fujihara H (2008) Asymmetric Suzuki-Miyaura coupling reactions catalyzed by chiral palladium nanoparticles at room temperature. Angew Chem Int Ed 120(36):7023–7025

49. Crespo-Quesada M, Yarulin A, Jin M, Xia Y, Kiwi-Minsker L (2011) Structure sensitivity of alkynol hydrogenation on shape-and size-controlled palladium nanocrystals: which sites are most active and selective? J Am Chem Soc 133(32):12787–12794

50. Mostafa S, Behafarid F, Croy JR, Ono LK, Li L, Yang JC, Frenkel AI, Cuenya BR (2010) Shape-dependent catalytic properties of Pt nanoparticles. J Am Chem Soc 132(44):15714–15719

51. Polshettiwar V, Luque R, Fihri A, Zhu H, Bouhrara M, Basset J-M (2011) Magnetically recoverable nanocatalysts. Chem Rev 111(5):3036–3075

52. Zhu Y, Stubbs LP, Ho F, Liu R, Ship CP, Maguire JA, Hosmane NS (2010) Magnetic nanocomposites: a new perspective in catalysis. Chemcatchem 2(4):365–374

53. Shylesh S, Schünemann V, Thiel WR (2010) Magnetically separable nanocatalysts: bridges between homogeneous and heterogeneous catalysis. Angew Chem Int Ed 49(20):3428–3459

54. Zhang X, Corma A (2008) Supported gold (III) catalysts for highly efficient three-component coupling reactions. Angew Chem Int Ed 120(23):4430–4433

55. Wang F, Ueda W, Xu J (2012) Detection and measurement of surface electron transfer on reduced molybdenum oxides (MoOx) and catalytic activities of Au/MoOx. Angew Chem Int Ed 51(16):3883–3887

56. Tsunoyama H, Ichikuni N, Sakurai H, Tsukuda T (2009) Effect of electronic structures of Au clusters stabilized by poly (N-vinyl-2-pyrrolidone) on aerobic oxidation catalysis. J Am Chem Soc 131(20):7086–7093

57. Cho SH, Kim JY, Kwak J, Chang S (2011) Recent advances in the transition metal-catalyzed twofold oxidative C-H bond activation strategy for C–C and C–N bond formation. Chem Soc Rev 40(10):5068–5083

58. Xu J, Ying JY (2006) A highly active and selective nanocomposite catalyst for C7+ paraffin isomerization. Angew Chem Int Ed 45(40):6700–6704

59. Haruta M, Yamada N, Kobayashi T, Iijima S (1989) Gold catalysts prepared by coprecipitation for low-temperature oxidation of hydrogen and of carbon monoxide. J Catal 115(2):301–309

60. Zhou K, Wang X, Sun X, Peng Q, Li Y (2005) Enhanced catalytic activity of ceria nanorods from well-defined reactive crystal planes. J Catal 229(1):206–212

61. Li Z, Yu R, Huang J, Shi Y, Zhang D, Zhong X, Wang D, Wu Y, Li Y (2015) Platinum-nickel frame within metal-organic framework fabricated in situ for hydrogen enrichment and molecular sieving. Nat Commun 6

62. Kamegawa T, Matsuura S, Seto H, Yamashita H (2013) A Visible-Light-Harvesting Assembly with a Sulfocalixarene linker between dyes and a Pt–TiO$_2$ photocatalyst. Angew Chem Int Ed 52(3):916–919

63. Li R, Zhang F, Wang D, Yang J, Li M, Zhu J, Zhou X, Han H, Li C (2013) Spatial separation of photogenerated electrons and holes among 010 and 110 crystal facets of BiVO$_4$. Nat Commun 4:1432

64. Liu H, Song C, Zhang L, Zhang J, Wang H, Wilkinson DP (2006) A review of anode catalysis in the direct methanol fuel cell. J Power Sources 155(2):95–110

65. Wu J, Yuan XZ, Martin JJ, Wang H, Zhang J, Shen J, Wu S, Merida W (2008) A review of PEM fuel cell durability: degradation mechanisms and mitigation strategies. J Power Sources 184(1):104–119

66. Cheng X, Shi Z, Glass N, Zhang L, Zhang J, Song D, Liu Z-S, Wang H, Shen J (2007) A review of PEM hydrogen fuel cell contamination: impacts, mechanisms, and mitigation. J Power Sources 165(2):739–756

67. Kim MG, Cho J (2009) Reversible and high-capacity nanostructured electrode materials for Li-Ion batteries. Adv Funct Mater 19(10):1497–1514

68. Kang B, Ceder G (2009) Battery materials for ultrafast charging and discharging. Nature 458(7235):190–193

69. Schätz A, Reiser O, Stark WJ (2010) Nanoparticles as semi-heterogeneous catalyst supports. Chem-Eur J 16(30):8950–8967

70. Díaz U, Brunel D, Corma A (2013) Catalysis using multifunctional organosiliceous hybrid materials. Chem Soc Rev 42(9):4083–4097

71. Rudolph M, Hashmi ASK (2012) Gold catalysis in total synthesis—an update. Chem Soc Rev 41(6):2448–2462

72. Haruta M (2004) Gold as a novel catalyst in the 21st century: preparation, working mechanism and applications. Gold Bull 37(1–2):27–36

73. Green IX, Tang W, Neurock M, Yates JT (2011) Spectroscopic observation of dual catalytic sites during oxidation of CO on a Au/TiO$_2$ catalyst. Science 333(6043):736–739

74. Edwards PP, Thomas JM (2007) Gold in a metallic divided state—from faraday to present-day nanoscience. Angew Chem Int Ed 46(29):5480–5486

75. Freund HJ, Meijer G, Scheffler M, Schlögl R, Wolf M (2011) CO oxidation as a prototypical reaction for heterogeneous processes. Angew Chem Int Ed 50(43):10064–10094

76. Tao FF, Salmeron M (2011) In situ studies of chemistry and structure of materials in reactive environments. Science 331(6014):171–174

77. Campbell CT (2004) The active site in nanoparticle gold catalysis. Science 306(5694):234–235

78. Sinfelt JH (1987) Structure of bimetallic clusters. Acc Chem Res 20(4):134–139

79. Heck RM, Farrauto RJ, Gulati ST (2012) Catalytic air pollution control: commercial technology. Wiley, Hoboken

80. Zhang J, Sasaki K, Sutter E, Adzic R (2007) Stabilization of platinum oxygen-reduction electrocatalysts using gold clusters. Science 315(5809):220–222

81. Rodriguez J (1996) Physical and chemical properties of bimetallic surfaces. Surf Sci Rep 24(7):223–287

82. Hammer B, Nørskov JK (2000) Theoretical surface science and catalysis—calculations and concepts. Adv Catal 45:71–129

83. Wang A-Q, Chang C-M, Mou C-Y (2005) Evolution of catalytic activity of Au–Ag bimetallic nanoparticles on mesoporous support for CO oxidation. J Phys Chem B 109(40):18860–18867

84. Somorjai GA, Park JY (2008) Molecular factors of catalytic selectivity. Angew Chem Int Ed 47(48):9212–9228

85. Yu W, Porosoff MD, Chen JG (2012) Review of Pt-based bimetallic catalysis: from model surfaces to supported catalysts. Chem Rev 112(11):5780–5817

86. Stamenkovic VR, Fowler B, Mun BS, Wang GF, Ross PN, Lucas CA, Markovic NM (2007) Improved oxygen reduction activity on Pt3Ni(111) via increased surface site availability. Science 315(5811):493–497

87. Wang C, Chi M, Wang G, Van der Vliet D, Li D, More K, Wang HH, Schlueter JA, Markovic NM, Stamenkovic VR (2011) Correlation between surface chemistry and electrocatalytic properties of monodisperse PtxNi1−x nanoparticles. Adv Funct Mater 21(1):147–152

88. Strasser P, Koh S, Anniyev T, Greeley J, More K, Yu C, Liu Z, Kaya S, Nordlund D, Ogasawara H (2010) Lattice-strain control of the activity in dealloyed core-shell fuel cell catalysts. Nat Chem 2(6):454–460

89. Chen C, Kang Y, Huo Z, Zhu Z, Huang W, Xin HL, Snyder JD, Li D, Herron JA, Mavrikakis M, Chi M, More KL, Li Y, Markovic NM, Somorjai GA, Yang P, Stamenkovic VR (2014) Highly crystalline multimetallic nanoframes with three-dimensional electrocatalytic surfaces. Science

90. Chen M, Kumar D, Yi C-W, Goodman DW (2005) The promotional effect of gold in catalysis by palladium-gold. Science 310(5746):291–293

91. Ferrin P, Mavrikakis M (2009) Structure sensitivity of methanol electrooxidation on transition metals. J Am Chem Soc 131(40):14381–14389

92. Niu Z, Wang D, Yu R, Peng Q, Li Y (2012) Highly branched Pt–Ni nanocrystals enclosed by stepped surface for methanol oxidation. Chem Sci 3(6):1925–1929

93. Quan Z, Wang Y, Fang J (2012) High-index faceted noble metal nanocrystals. Acc Chem Res 46(2):191–202

94. Yin A-X, Min X-Q, Zhu W, Liu W-C, Zhang Y-W, Yan C-H (2012) Pt–Cu and Pt–Pd–Cu concave nanocubes with high-index facets and superior electrocatalytic activity. Chem A Eu J 18(3):777–782

95. Wang D, Li Y (2011) Bimetallic nanocrystals: liquid-phase synthesis and catalytic applications. Adv Mater 23(9):1044–1060

96. Wu JB, Gross A, Yang H (2011) Shape and composition-controlled platinum alloy nanocrystals using carbon monoxide as reducing agent. Nano Lett 11(2):798–802

97. Zhang J, Yang HZ, Fang JY, Zou SZ (2010) Synthesis and oxygen reduction activity of shape-controlled Pt(3)Ni nanopolyhedra. Nano Lett 10(2):638–644

98. Wu J, Zhang J, Peng Z, Yang S, Wagner FT, Yang H (2010) Truncated octahedral Pt3Ni oxygen reduction reaction electrocatalysts. J Am Chem Soc 132(14):4984–4985

99. Wu J, Qi L, You H, Gross A, Li J, Yang H (2012) Icosahedral platinum alloy nanocrystals with enhanced electrocatalytic activities. J Am Chem Soc 134(29):11880–11883

100. Cui C, Gan L, Li H-H, Yu S-H, Heggen M, Strasser P (2012) Octahedral PtNi nanoparticle catalysts: exceptional oxygen reduction activity by tuning the alloy particle surface composition. Nano Lett 12(11):5885–5889

101. Wang D, Zhao P, Li Y (2011) General preparation for Pt-based alloy nanoporous nanoparticles as potential nanocatalysts. Sci Rep-Uk 1 47(1):1–5

102. Snyder J, McCue I, Livi K, Erlebacher J (2012) Structure/processing/properties relationships in nanoporous nanoparticles as applied to catalysis of the cathodic oxygen reduction reaction. J Am Chem Soc 134(20):8633–8645

103. Cui C, Gan L, Heggen M, Rudi S, Strasser P (2013) Compositional segregation in shaped Pt alloy nanoparticles and their structural behaviour during electrocatalysis. Nat Mater 12(8):765–771

104. Xia Y, Xiong Y, Lim B, Skrabalak SE (2009) Shape-controlled synthesis of metal nanocrystals: simple chemistry meets complex physics? Angew Chem Int Ed 48(1):60–103

105. Wu B, Zheng N (2013) Surface and interface control of noble metal nanocrystals for catalytic and electrocatalytic applications. Nano Today 8(2):168–197
106. Cong H, Porco JA Jr (2011) Chemical synthesis of complex molecules using nanoparticle catalysis. Acs Catal 2(1):65–70
107. Witham CA, Huang W, Tsung C-K, Kuhn JN, Somorjai GA, Toste FD (2009) Converting homogeneous to heterogeneous in electrophilic catalysis using monodisperse metal nanoparticles. Nat Chem 2(1):36–41
108. Oliver-Meseguer J, Cabrero-Antonino JR, Domínguez I, Leyva-Pérez A, Corma A (2012) Small gold clusters formed in solution give reaction turnover numbers of 107 at room temperature. Science 338(6113):1452–1455
109. Pun D, Diao T, Stahl SS (2013) Aerobic dehydrogenation of cyclohexanone to phenol catalyzed by Pd (TFA) 2/2-dimethylaminopyridine: evidence for the role of Pd-nanoparticles. J Am Chem Soc 135(22):8213–8221

Chapter 2
Controllable Synthesis of Water-Soluble Pt–Ni Alloys and the Study of Their Catalytic Properties

2.1 Introduction

Noble metal nanoparticles (NPs) (i.e., Pt, Pd, [1–3] etc.) have been widely used as catalysts in many applications such as organic reactions [4], fuel cells [5], and so on, while their properties are closely bound up to well-defined sizes, compositions, and structures [6]. To improve atomic efficiency and sustainability, researchers have studied the Pt-based bimetallic nanostructures in which Pt is partially replaced by cheap 3D-transition metals (i.e., Fe, Co, Ni, etc.) [7, 8]. Considering the limited resource and rising cost of Pt, this bimetallic nanostructures will not only efficiently reduce the consumption of Pt, but also exhibit superior catalytic performances than the single component, possibly due to the different geometric and electronic structures brought by the introduction of the second metal [9, 10].

The hydrogenation reactions of benzalacetone, styrene, and nitrobenzene are important industrial processes [11–13] in which homogeneous catalysts have been widely used, but the recycling of catalysts remains difficult. Owning high selectivity for the desired product without sacrificing activity, heterogeneous catalysts [14] have attracted much attention for the ease of their separation and recycling [15, 16]. Besides, as a favorable heterogeneous catalyst, bimetallic catalysts could have optimized activity, selectivity, and stability due to the possible synergistic effects.

So far, most of the alloys that consist of Pt and 3D-transition metals (i.e., Fe, Co, Ni) have been produced with hydrophobic solvents while few cases are reported to synthesize the materials in a hydrophilic system. In this chapter, we developed a general method for the shape and composition controlled synthesis of a series of water-soluble Pt–Ni alloy utilizing poly(vinylpyrrolidone) as a surfactant. As a complement to the hydrophilic system, the water-soluble catalysts produced by this method are also compatible with hydrophobic solvents through a ligands exchange process. The shapes of octahedral, truncated octahedral, and cubic Pt–Ni alloy were controlled by varying the growth inhibition agents such

© Springer-Verlag Berlin Heidelberg 2016
Y. Wu, *Controlled Synthesis of Pt–Ni Bimetallic Catalysts
and Study of Their Catalytic Properties*, Springer Theses,
DOI 10.1007/978-3-662-49847-7_2

as benzoic acid, aniline, carbon monoxide, and potassium bromide. In a preliminary structure-activity study, we found that the rates of hydrogenation reactions were significantly affected by the shape and composition of as-prepared nanocrystals. Besides, we found that the octahedral $PtNi_2$ alloy exhibited superior reactivity toward both the hydrogenation of the C=C and N=O bonds. On the other hand, the Pt–Ni alloy nanocrystal has much higher catalytic activity than its monometallic counterparts, endowing the hydrogenation of the C=C and N=O bonds a great selectivity and activity at room temperature under atmospheric pressure.

2.2 Experimental Section

2.2.1 Chemicals and Instruments

Chemicals: Analytical grade benzyl alcohol, aniline benzaldehyde, benzoic acid, and salicylic acid were obtained from Beijing Chemical Reagents, P.R. China. $Pt(acac)_2$ (99 %), $Ni(acac)_2$ (99 %), and PVP (MW = 8000, AR) were purchased from Alfa Aesar. Oleylamine (80–90 %) was purchased from Acros., Pt/C (10 wt%) was purchased from Sigma Aldrich, activated carbon (Vulcan XC-72) was purchased from Carbot. All of the chemicals were used without further purification.

Instruments: XRD patterns were recorded by Rigaku D/Max 2500Pc X-ray powder diffractometer with CuKa radiation ($\lambda = 1.5418$ Å). TEM images and HRTEM images were recorded by a Hitachi H-800 transmission electron microscope working at 100 kV and a FEI Tecnai G2 F20 S-Twin high-resolution transmission electron microscope working at 200 kV, respectively. Electrochemical measurements were conducted on CHI 660D electrochemical analyzer. X-ray photoelectron spectroscopy (XPS) experiments were performed on a ULVAC PHI Quantera microprobe. Binding energies (BE) were calibrated by setting the measured BE of C1s–284.8 eV. The catalytic reaction results were measured by gas chromatography (GC) (SP-6890) and gas chromatography–mass spectroscopy (GC–MS) (ITQ 700/900/1100) and ^1H NMR. The NMR spectroscopy was conducted on a JEOL JNM-ECX 400 MHz instrument.

2.2.2 Experimental Methods

Preparation of Pt–Ni Octahedral Nanocrystals. In a typical synthesis of $PtNi_2$ octahedral nanocrystals, $Pt(acac)_2$, (8.0 mg), poly(vinylpyrrolidone) (PVP, MW = 8000), (80.0 mg), $Ni(acac)_2$ (10.0 mg), benzoic acid (50 mg), and benzylalcohol, (5 ml) were added to a 12-ml Teflon-lined stainless-steel autoclave, followed by 5–10 min vigorous stirring at room temperature. The sealed vessel was then heated at 150 °C for a 12 h. When it was cooled down to room temperature, the products were first precipitated by excess acetone, separated via centrifugation, and further purified by an ethanol-acetone mixture for 3 times.

Preparation of Pt–Ni Truncated Octahedral Nanocrystals. In a typical synthesis of PtNi$_2$ truncated octahedral nanocrystals, Pt(acac)$_2$, (8.0 mg), poly(vinylpyrrolidone) (PVP, MW = 8000), (80.0 mg), Ni(acac)$_2$ (10.0 mg), and aniline (0.1 ml), benzylalcohol, (5 ml) were added to a 12-ml Teflon-lined stainless-steel autoclave, followed by 5–10 min vigorous stirring at room temperature. The sealed vessel was then heated at 150 °C for a 12 h. When it was cooled down to room temperature, the products were first precipitated by excess acetone, separated via centrifugation, and further purified by an ethanol-acetone mixture for 3 times.

Preparation of Pt–Ni Cubic Nanocrystals. In a typical synthesis of PtNi$_2$ cubic nanocrystals, Pt(acac)$_2$, (8.0 mg), poly(vinylpyrrolidone) (PVP, MW = 8000), (80.0 mg), Ni(acac)$_2$, (10.0 mg) and potassium bromide (50.0 mg), and benzylalcohol, (5 ml) were added to a 10-ml round-bottom flask equipped with a magnetic stirrer, followed by 5–10 min vigorous stirring at room temperature. After aerating the reaction flask full of carbon monoxide, the flask was then immersed in an oil bath at 150 °C for 4 h under a carbon monoxide balloon. When it was cooled down to room temperature, the products were first precipitated by excess acetone, separated via centrifugation, and further purified by an ethanol-acetone mixture for 3 times.

Preparation of Ni Octahedral and Truncated Octahedral Nanocrystals. In a typical synthesis of octahedral Ni nanocrystals, poly(vinylpyrrolidone) [PVP, MW = 8000, (80.0 mg)], Ni(acac)$_2$ (10 mg), benzaldehyde (0.5 mL), benzoic acid (50.0 mg), and benzyl alcohol (5 mL) were added to a 12-mL Teflon-lined stainless-steel autoclave, followed by 5–10 min vigorous stirring at room temperature. The sealed vessel was then heated at 200 °C for 12 h. When it was cooled down to room temperature, the black nanoparticles were precipitated by acetone, separated via centrifugation, and further purified by an ethanol-acetone mixture for 3 times. The Ni-truncated octahedron was synthesized using the above procedure with aniline (0.1 mL) instead of benzaldehyde and benzoic acid.

Typical Procedure for the Hydrogenation of Benzalacetone, Styrene, and Nitrobenzene. To a 10-mL round-bottom flask was charged a solution of the respective substrate (0.5 mmol benzalacetone, or styrene, or nitrobenzene) in THF (2.5 mL) and the Pt–Ni catalyst (0.01 mmol, 2 mol%). After being purged with H$_2$, the reaction mixture was stirred in an oil bath at 25 °C under a H$_2$ balloon. The product was purified by column chromatography and subsequently characterized by ^{1}H NMR.

Preparation of Oleylamine-Capped Pt–Ni Alloy. In the synthesis of oleylamine-capped Pt–Ni alloy, Pt–Ni catalyst (10.0 mg) and oleylamine (0.1 ml) were mixed together with toluene (5 ml) in a 10-ml round-bottom flask. The resulting mixture was refluxed for 12 h at 100 °C. The products were precipitated by adding ethanol, separated via centrifugation, and further purified by an ethanol-cyclohexane mixture for 3 times.

Preparation of Clean Pt–Ni/C Catalysts. In the synthesis of clean Pt–Ni/C catalysts, activated carbon (100.0 mg) were mixed together with 30 ml of ethanol in a 100 ml beaker, followed by sonication for 30 min to ensure the activated carbon is adequately dispersed. To this beaker Pt–Ni catalyst (10.0 mg) was then added and then sonicated for 1 h. The resulting black mixture was transferred into a centrifuge tube, separated via centrifugation, and further purified by ethanol-cyclohexane mixture for 3 times. The as-prepared solid was dispersed in 40 ml of acetic acid and heated at 70 °C for 10 h with vigorous stirring. When it was cooled down to room temperature, the resulting black mixture was separated via centrifugation with the dumping of the clear liquid, followed by washing the black solid with ethanol for 3 times and dried in an oven.

Catalyst Recovery. The Pt–Ni catalysts were separated from the reaction mixture by a magnet. After being washed with ethanol for 3 times, the recovered catalysts were directly used in the next cycle of the reaction without further manipulations.

2.3 Results and Discussion

2.3.1 Structural Characterization

The representative transmission electron microscopy (TEM) images (Fig. 2.1) showed that the as-prepared Pt–Ni alloys were all uniform in terms of the narrow size distribution, shape, and exclusively bound by well-defined {111} or {100} facets. The high-resolution TEM (HRTEM) images and fast Fourier transform (FFT) images (Fig. 2.1b, f, j) on single NP both confirmed their single-crystal-line nature. Continuous lattice fringes measurements with an interplanar spacing of 0.213 and 0.186 nm could be assigned to the corresponding {111} facets and {100} facets of fcc Pt–Ni alloy, respectively.

Pt–Ni Octahedrons. The octahedral Pt–Ni particles have an average distance of 11.8 ± 1.2 nm from one apex to the opposite apex and consist of only eight equivalent {111} facets. The high-angle annular dark-field scanning transmission electron microscope (HAADF-STEM) micrographs showed that all the surfaces of the octahedrons were smooth without obvious defect throughout the nanocrystal. The corresponding elemental maps of HAADF-STEM (Fig. 2.1c) showed that both Pt and Ni were distributed evenly throughout each individual $PtNi_2$ alloy octahedral nanocrystal. To realize the formation of Pt–Ni octahedrons preferably, this chapter taken a study on the effect of the amounts of benzoic acid on the shape. Indeed, the yields of octahedrons significantly increased from 40 to 95 % with the amounts of benzoic acid from 0 to 0.04 mmol. In contrast, the yield was almost unaffected by further increasing the amount of benzoic acid. Herein, the usage of benzoic acid was believed to play an important role in the formation of Pt–Ni octahedrons (Fig. 2.2).

Pt–Ni Truncated Octahedrons. By replacing benzoic acid with aniline, we could obtain a series of Pt–Ni alloy with truncated octahedral shape. The truncated

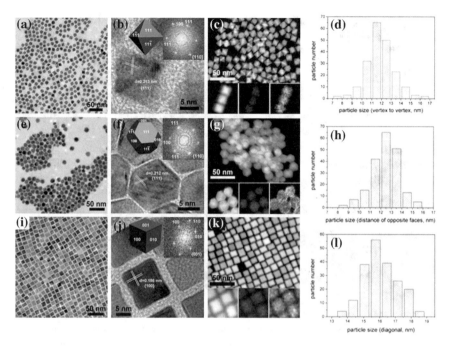

Fig. 2.1 a TEM, **b** HRTEM (*Top-right* and *top-middle* insets show the corresponding FFT pattern and the ideal structure model), and **c** HAADF-STEM images of PtNi$_2$ octahedrons, corresponding element maps showing the distribution of Pt (*yellow*) and Ni (*red*). **d** Size distribution of octahedral particles (average size: 11.8 ± 1.2 nm, 95 % octahedrons and 5 % irregular shapes). **e** TEM, **f** HRTEM, and **g** HAADF-STEM images of PtNi$_2$-truncated octahedrons. **h** Size distribution of truncated octahedral particles (average size: 12.5 ± 1.1 nm, 90 % truncated octahedrons and 10 % irregular shapes). **i** TEM, **j** HRTEM, and **k** HAADF-STEM images of PtNi$_2$ cubes. **l** Size distribution of cubic particles (average size: 16.1 ± 1.7 nm, 95 % octahedrons and 5 % irregular shapes). Reprinted with the permission from Ref. [17]. Copyright 2012 American Chemical Society

octahedrons consisted of both {100} and {111} facets and the average distance of two opposite faces was 12.5 ± 1.1 nm (Fig. 2.1e). In our synthetic method, the benzoic alcohol was used as solvent and reductant; however, the addition of aniline was believed to add as coreductant and affect the reaction kinetics of the reduction of the nanocrystals [18]. The sizes of truncated octahedral Pt–Ni alloy could be tuned from 16.0 to 12.5 nm, 7.2 nm, and 4.8 nm by varying the quantity of aniline. Herein, the nucleation process could be facilitated by the addition of aniline, and the fleetly and largely increase of these nuclei densities significantly helps to reduce the size of the truncated octahedrons while the aniline was added as a strong reductant. The Fig. 2.3 showed that a truncated octahedron can be generated by cutting off six vertices from octahedron, where a was the distance from the vertex to the body center, then b was the height of each square pyramid, so the degree of truncation can be defined as b/a.

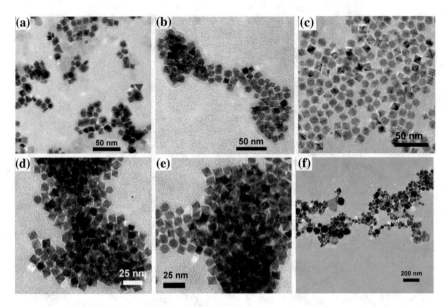

Fig. 2.2 TEM images show the effect of benzoic acid amounts on the formation of PtNi$_2$ octahedral nanocrystals. The amounts of benzoic acid used were **a** without adding any benzoic acid (40 % octahedrons and 60 % irregular shapes), **b** 0.005 g (0.04 mmol) (70 % octahedrons and 30 % irregular shapes) and **c** 0.25 g (2.0 mmol) (95 % octahedrons and 5 % irregular shapes). TEM images of PtNi$_2$ octahedrons prepared by replacing benzoic acid with **d** salicylic acid and **e** sodium benzoate. **f** TEM image of nanocrystals collected from the reaction with the same conditions used in the synthesis of octahedral Ni but without adding any benzoic acid (15 % octahedrons and 85 % irregular shapes). Reprinted with the permission from Ref. [17]. Copyright 2012 American Chemical Society

Fig. 2.3 Schematic structure of truncated octahedron. Reprinted with the permission from Ref. [17]. Copyright 2012 American Chemical Society

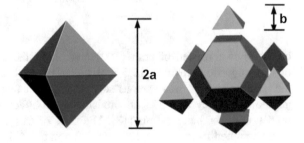

Figure 2.4 showed that the degree of truncation of the as-prepared truncated octahedron depended on the molar ratio between benzoic acid and aniline, a higher molar ratio yielded products with a much greater degree of truncation.

Unlike the typical octahedral Pt–Ni alloy, the elemental map of the truncated octahedral Pt–Ni alloy showed that a greater quantity of Pt segregated at the edges of the particles and Ni appeared more centrally located (Fig. 2.1g), which was

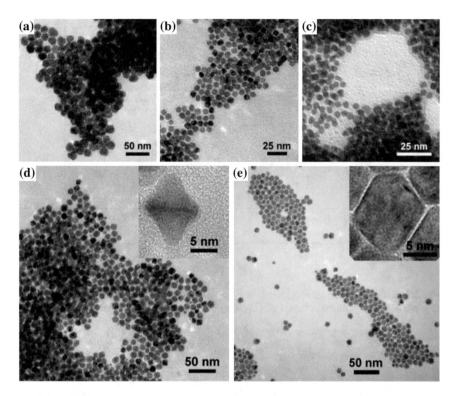

Fig. 2.4 TEM images show the effect of aniline amounts on the shape of Pt–Ni alloy. The amounts of aniline and benzoic alcohol used were **a** 20 μl aniline + 5 ml benzoic alcohol (average distance between two opposite faces is 16.0 nm), **b** 500 μl aniline + 5 ml benzoic alcohol (average distance between two opposite faces is 7.3 nm) and **c** 5 ml aniline (average distance between two opposite faces is 7.3 nm). TEM images of Pt–Ni alloy prepared by varying the molar ratio between benzoic acid and aniline **d** 0.15 g benzoic acid + 50 μl aniline (truncation degree is 0.28) **e** 0.017 g benzoic acid + 300 μl aniline (truncation degree is 0.32). Reprinted with the permission from Ref. [17]. Copyright 2012 American Chemical Society

similar to the core-shell structure. This structure of uneven distribution was further verified by the following line scan of an individual particle (Fig. 2.5).

Pt–Ni Cube. Previous studies have revealed that the carbon monoxide and bromide species could be selectively adsorbed onto Pt {100} crystal faces and induce the formation of nanocubes [19, 20]. Here, we successfully extended this method to the preparation of well-defined Pt–Ni alloy nanocubes. The as-prepared cubic particles had an average diagonal length of 16.1 ± 1.7 nm and were surrounded by six equivalent {100} facets. To better understand the role of CO and KBr in the formation of our Pt–Ni nanocubes, we carried out a series of control experiments. Figure 2.6 showed that the yield of nanocube decreased from 95 to 60 % with the absence of KBr. However, in the absence of CO, mixed morphologies (30 % nanorods, 40 % nanocubes, and 30 % irregular shapes) were observed.

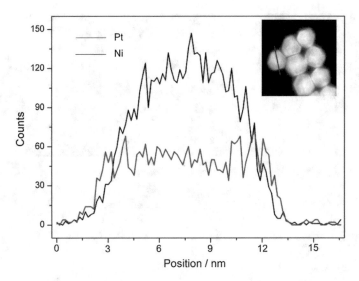

Fig. 2.5 Cross-sectional compositional line profiles of a truncated octahedral PtNi$_2$ alloy. Reprinted with the permission from Ref. [17]. Copyright 2012 American Chemical Society

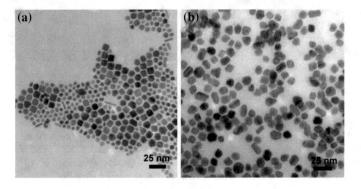

Fig. 2.6 TEM images of nanocrystals collected from the reaction with the same conditions used in the synthesis of cubic PtNi$_2$ but **a** without adding KBr (60 % octahedrons and 40 % irregular shapes) or **b** without adding CO (30 % nanorods, 40 % nanocubes, and 30 % irregular shapes). Reprinted with the permission from Ref. [17]. Copyright 2012 American Chemical Society

Ni Nanocrystals. This general method could also be applied to the synthesis of Ni octahedrons with an average size of 40 ± 4.5 nm, albeit at an elevated temperature of 200 °C when benzaldehyde was used as an extra reductant (Fig. 2.7a). Similarly, the truncated octahedral Ni with an average size of 25.2 ± 4.3 nm was synthesized when aniline replaced benzoic acid and benzaldehyde. Unfortunately, this route could not be extended to the synthesis of Ni nanocubes. It is intriguing that these magnetic Ni octahedrons spontaneously assembled into large-scale one-dimensional nanostructures on copper grid after ethanol was evaporated (Fig. 2.7b–d). The self-assemble behavior of the magnetic nanoparticles was

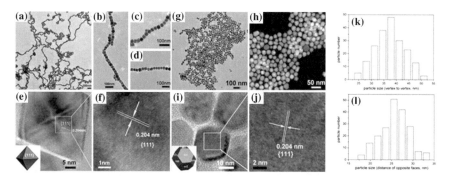

Fig. 2.7 a TEM image of Ni octahedrons **b–d** Magnified TEM images of Ni 1D structures with different orientation **e** HRTEM image of Ni octahedron; the inset is the scheme of the structure of octahedron; **f** Magnified images of the lattice fringe of the Ni octahedron. **g** TEM, **h** HAADF-STEM, and **i** images of Ni-truncated octahedrons; **j** magnified images of the lattice fringe of the Ni-truncated octahedron size distribution of **k** octahedral Ni nanoparticles (average size: 40.0 ± 5.1 nm, 90 % octahedrons, and 5 % irregular shapes) and **l** truncated octahedral Ni nanoparticles (average size: 25.2 ± 4.3 nm, 80 % octahedrons, 15 % nanoplate, and 5 % irregular shapes). Reprinted with the permission from Ref. [17]. Copyright 2012 American Chemical Society

corresponding to the Stoner–Wohlfarth model, [21] which was believed to be a result of their magnetic anisotropy [22–24]. The HRTEM of as-synthesized Ni NPs showed lattice fringes of 0.204 nm, which could be assigned to {111} planes of Ni in face-centered cubic (fcc) phase (Fig. 2.7e, f, i, j).

Importantly, our method is found to be generally applicable to the synthesis of other water-soluble bimetallic nanocrystals such as RhNi, PdNi, PdCu, and PtCu with excellent monodispersity and crystallization (Fig. 2.8).

The Fig. 2.9 showed that the X-ray diffraction (XRD) patterns of all these Pt–Ni nanocrystals could be indexed to {111}, {200}, and {220} diffractions of fcc metal with the peak positioned between those of Pt (JCPD-04–0802) and Ni (JCPD-04–0850) standard peak. We noted that the peaks continuously shifted from the Pt standard peaks to Ni standard ones when the content of Ni increases. Besides, the composition of the Pt–Ni alloy could be effectively controlled by choice of ratios between the two precursors, which was further verified by Energy-dispersive X-ray spectroscopy (EDX) and ICP-AES.

It is believed that the formation of the bimetallic alloy in the synthesis system developed in this chapter is the result of a noble-metal-induced-reduction process [25]. That is the Pt (0) species would perform to facilitate the reduction of Ni (II) to Ni (0) in the growth of Pt–Ni nanocrystals.

2.3.2 Surface Modification of Water-Soluble Pt–Ni Alloy

It is important to note two features of our method: the use of alcohol with high boiling point as solvents, phenyl organic molecules as growth inhibition agents,

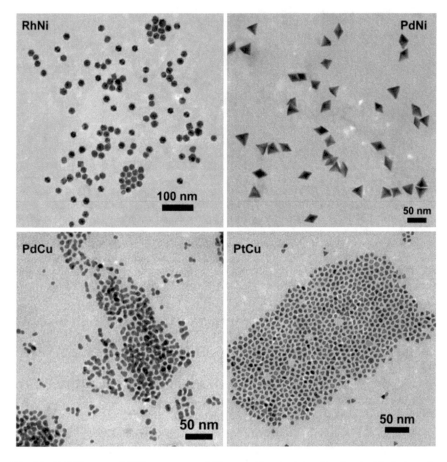

Fig. 2.8 TEM images of different water-soluble nanoalloy particles. Reprinted with the permission from Ref. [17]. Copyright 2012 American Chemical Society

and hydrophilic PVP as the capping agent. On the one hand, several groups have reported the shape control synthesis of Pt-based alloys, [26–28] which usually involved the use of reductants and growth inhibition agents such as $W(CO)_6$ [29] or CO [30]. In these methods, the use of reagents was mainly credited to the selective binding of CO [31] or bromide anion to Pt. Similarly, in our method, benzoic acid and aniline played a crucial role in the growth of Pt–Ni alloy, probably due to the selective adsorption effect that suppresses crystal growth in certain direction. However, the selective adsorption effect of phenyl organic molecules were well documented, such as on metal (Au, [32] Pt, [33] Ni [34]), on oxides [35] and on carbon materials [36]. This controlling effect of the phenyl organic molecules was substantiated by the fact that substituting benzoic acid with salicylic acid or sodium benzoate also resulted in the controlled synthesis of the Pt–Ni alloy.

On the other hand, the commonly used surfactants such as oleylamine, oleic acid in hydrophobic solvents usually resulted in catalysts of low activity and

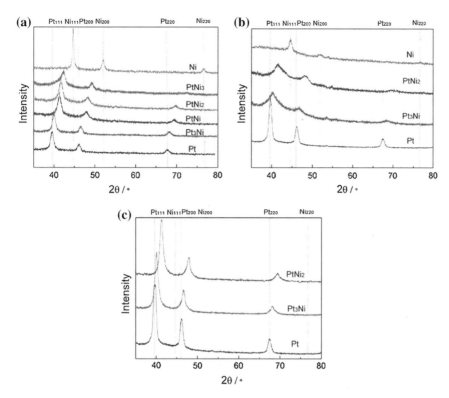

Fig. 2.9 XRD patterns of as-prepared Pt, Pt–Ni and Ni **a** octahedral nanocrystals, **b** truncated octahedral nanocrystals and **c** cubic nanocrystals. Reprinted with the permission from Ref. [17]. Copyright 2012 American Chemical Society

selectivity, possibly due to the strong chemical bonds on the catalysts' surface [37]. Herein, there must be a post treatment for these catalysts to obtain a desired higher catalytic activity while our Pt–Ni nanocrystals prepared in the presence of PVP could be used without any post treatment. Furthermore, we can modify the surface of these water-soluble nanoparticles with hydrophobic species by a simple process of exchanging the ligands, which could make the prepared catalysts a great ability to adapt to the complex reaction situation (Fig. 2.10).

2.3.3 Catalytic Performance of Pt–Ni Alloy in the Hydrogenation Reactions

We chose hydrogenation of benzalacetone, styrene, and nitrobenzene as the laboratory test reactions to probe our nanocrystals' structure–activity relationship, which would shed light on further application of the water-soluble Pt–Ni alloy. Notably, bimetallic catalysts are inspiring for possible synergistic effects that offer a higher catalytic activity for these hydrogenation (Fig. 2.11).

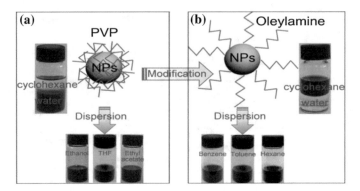

Fig. 2.10 Surface modification of Pt–Ni alloy. Reprinted with the permission from Ref. [17]. Copyright 2012 American Chemical Society

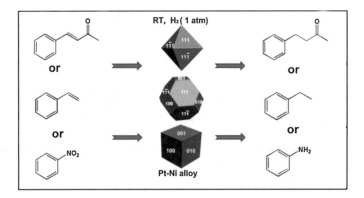

Fig. 2.11 Hydrogenation of Benzalacetone, Styrene, and Nitrobenzene. Reprinted with the permission from Ref. [17]. Copyright 2012 American Chemical Society

We first studied the effects of solvents, which showed that the use of hydrophilic solvents such as ethanol and ethyl acetate could also obtain excellent catalytic activity with the exception of THF. To this end, three types of catalyst [i,e., Ni, Pt, and Pt_xNi_{1-x} $(0 < x < 1)$] with comparable composition and different shapes were tested to ensure a meaningful comparison. It is interesting to note that the catalyst activity in hydrogenation of benzalacetone increase with the increased percentage of {111} facets [38]. Next, we examined the composition-activity dependence using octahedral Pt–Ni catalysts of different Pt/Ni ratios. Overall, bimetallic nanoalloy catalysts are more active than their monometallic counterparts, in the order of $PtNi_2 > PtNi \approx PtNi_3 > Pt_3Ni > PtNi_{10} > Pt > Ni$. The same trend was also observed in the hydrogenation of styrene with these Pt–Ni octahedrons. The Turnover frequencies (TOF) measured using octahedral $PtNi_2$ was up to 6-times higher than that measured using octahedral Pt catalyst. It was very satisfying to see that the best catalyst achieved 100 % conversion with nearly 100 %

Table 2.1 Solvent effect of catalytic reactions[a]

Entry	Substrate	Solvent	Temp. (°C)	t. (h)	Conversion[b] (%)	Selectivity[c] (%)
1	benzalacetone	CH3OH	25	2	90	>99
2	benzalacetone	AcOEt	25	2	95	>99
3[d]	benzalacetone	THF	0	2	76	97
4	styrene	CH3OH	25	2	99	>99
5	styrene	AcOEt	25	2	99	>99
6[d]	styrene	THF	0	2	85	>99

[a]Conditions: substrate (0.5 mmol), 2 mol% catalyst with respect to the substrate and THF solvent (2.5 ml) were stirred under hydrogenation balloon/room temperature conditions
[b]Determined by GC using n-tridecane as an internal standard
[c]Selectivity (%) = [(GC peak area of desired product)/(GC peak area of total products)] × 100
[d]Reaction temperature: 273.15 K

Table 2.2 Turnover Frequencies [TOFs (h − 1)] of the Hydrogenation of Benzalacetone Catalyzed by Various Catalysts[a]

Entry	Catalyst	Average size. (nm)	Composition[b]	TOFs [h − 1]
1	Pt octahedron	11.9	–	23
2	PtNi$_2$ octahedron	11.8	0.37:0.63	139
3	Ni octahedron[c]	40.0	–	–
4	PtNi$_2$ truncated octahedron	12.5	0.36:0.64	114
5	PtNi$_2$ cube	11.2	0.35:0.65	95

[a]Conditions: substrate (0.5 mmol), 2 mol% catalyst with respect to the substrate and THF solvent (2.5 mL) were stirred under hydrogenation balloon/room temperature conditions
[b]The compositions were measured by ICP-AES
[c]Considering the ratio between surface area and volume, the amounts of Ni was 4 mol% respect to the substrate

selectivity in 2 h at room temperature under atmospheric H_2 pressure, which could effect both hydrogenation reactions 0 °C albeit at a slightly decreased rate (Tables 2.1 and 2.2).

We then evaluated PtNi$_2$ octahedron in the hydrogenation reactions of nitrobenzene, a more challenging reaction and important industrial process. It was again to our delight that nitrobenzene was quantitatively converted to aniline in 24 h, and the nanocrystals are substantially more active than the monometallic Pt and Ni nanocrystals (Fig. 2.12). The oxidation state of octahedral PtNi$_2$ and octahedral Ni nanocrystals were measured by XPS. The surface information about PtNi$_2$ alloy revealed the Pt 4f7/2 and 4f5/2 had a binding energy of 71.3 and 74.6 eV, characteristic of Pt (0). The Ni 2p1/2 and 2p3/2 were 869.9 and 852.8 eV, respectively, which corresponded to the Ni (0) (Fig. 2.13). In contrast, the XPS results showed that the oxidation state of Ni on the surfaces of octahedral Ni nanocrystals was mainly corresponded to Ni(II), which indicated the surfaces of octahedral Ni nanocrystals were dominated by a thin layer of oxidized Ni. Furthermore, the low activity of octahedral Ni catalyst could be contributed to the inertness of this thin

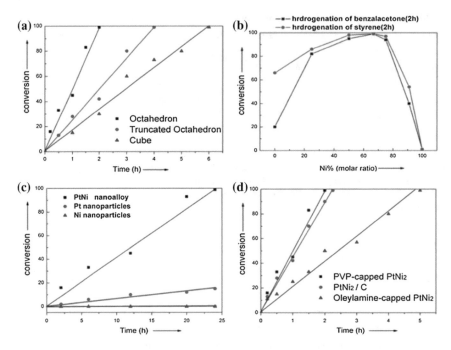

Fig. 2.12 **a** Conversion % as a function of time in hydrogenation of benzalacetone with PtNi$_2$ octahedral, truncated octahedral, and cubic nanocrystals. **b** Conversion % in a 2 h time of hydrogenation reactions of benzalacetone and styrene at r.t. under 1 atm H$_2$ pressure with octahedral NPs of different compositions. **c** Conversion % as a function of time in hydrogenation of nitrobenzene with PtNi$_2$, Ni, and Pt octahedrons. **d** Conversion % as a function of time in hydrogenation of benzalacetone with PVP-capped PtNi$_2$, "cleaned" PtNi$_2$/C and oleylamine-capped PtNi$_2$ octahedrons. Reprinted with the permission from Ref. [17]. Copyright 2012 American Chemical Society

layer of Ni (II). These date collectively suggest that shape, composition, and structure indeed play an important role in dictating the catalysts' activity and selectivity in the catalytic reactions.

One of the major goals in current heterogeneous catalyst research is to design nanocatalysts with good compatibility with complex reaction systems. We prepared the oleylamine-capped PtNi$_2$ nanocrystals through a ligand exchange process and a "cleaned" PtNi$_2$/C (10 % loading) catalyst by the washing with ethylic acid. In contrast, the oleylamine-capped nanocrystals are much less active toward hydrogenation of benzalacetone, while the PVP-capped nanocrystals show similar activity to that of the "cleaned" PtNi$_2$/C catalyst, underlying the superiority of PVP as the capping agent that exempted the catalyst from postsynthesis treatment [39]. Ecologically friendly catalysts, especially those involving precious metals, underscore the emphasis of green chemistry. Unlike the organic syntheses that are routinely performed in nonaqueous solution, nature carries out most of the chemical reactions in water [40]. Owing to their excellent bicompatibility and

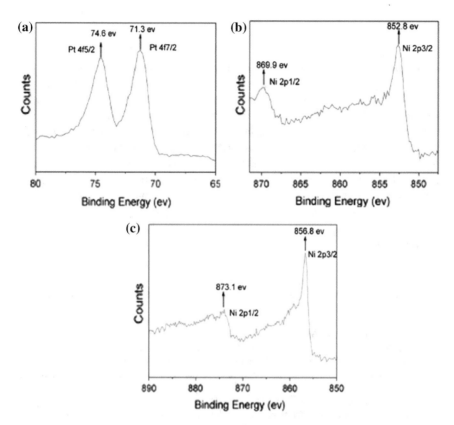

Fig. 2.13 XPS data for the octahedral PtNi$_2$ nanocrystals: high-resolution scans of the **a** Pt 4f and **b** Ni 2p.XPS data for the octahedral Ni nanocrystals: high-resolution scans of the **c** Ni 2p. Reprinted with the permission from Ref. [17]. Copyright 2012 American Chemical Society

hydrophily, the as-prepared water-soluble catalysts could find important applications in green chemistry and biology.

Importantly, the catalysts could be recycled at least five times without observable decay of activity. TEM study of all spent catalysts found no obvious aggregation or shape change, suggesting excellent stabilities of our nanocrystals under these reaction conditions (Table 2.3).

Table 2.3 Recycle of Catalysts

Entry	Catalyst	Temp. [°C]	t. [h]	Conversion [%]	Selectivity [%]
1	PtNi$_2$	25	2	>99	>99
2	1 recycle	25	2	>99	>99
3	2 recycle	25	2	>99	>99
4	3 recycle	25	2	>99	>99
5	4 recycle	25	2	>99	>99

2.4 Conclusions

In this chapter, we have developed a general method to synthesize water-soluble Pt, Pt_xNi_{1-x} ($0 < x < 1$), and Ni nanocrystals with uniformly controlled shapes and compositions. Through a series of model hydrogenation reactions, we performed a systemic evaluation of structure-activity relationship of the catalysts. Bimetallic nanoalloy catalysts are more active than their monometallic counterparts possibly due to a synergistic effect between the two metals. Using hydrophilic and green PVP as the capping agent could exempt the catalyst from postsynthesis treatment with great enhancement of catalyst's biocompatibility. Overall, this general method could be an effective tool for us to screen efficient catalyst in the future.

References

1. Huang XQ, Tang SH, Zhang HH, Zhou ZY, Zheng NF (2009) Controlled formation of concave tetrahedral/trigonal bipyramidal palladium nanocrystals. J Am Chem Soc 131(39):13916–13917
2. Niu ZQ, Peng Q, Gong M, Rong HP, Li YD (2011) Oleylamine-mediated shape evolution of palladium nanocrystals. Angew Chem Int Edit 50(28):6315–6319
3. Habas SE, Lee H, Radmilovic V, Somorjai GA, Yang P (2007) Shaping binary metal nanocrystals through epitaxial seeded growth. Nat Mater 6(9):692–697
4. Newton MA (2008) Dynamic adsorbate/reaction induced structural change of supported metal nanoparticles: heterogeneous catalysis and beyond. Chem Soc Rev 37(12):2644–2657
5. Bing YH, Liu HS, Zhang L, Ghosh D, Zhang JJ (2010) Nanostructured Pt-alloy electrocatalysts for PEM fuel cell oxygen reduction reaction. Chem Soc Rev 39(6):2184–2202
6. Daniel MC, Astruc D (2004) Gold nanoparticles: assembly, supramolecular chemistry, quantum-size-related properties, and applications toward biology, catalysis, and nanotechnology. Chem Rev 104(1):293–346
7. Stamenkovic VR, Fowler B, Mun BS, Wang GF, Ross PN, Lucas CA, Markovic NM (2007) Improved oxygen reduction activity on $Pt_3Ni(111)$ via increased surface site availability. Science 315(5811):493–497. doi:10.1126/science.1135941
8. Wu JB, Zhang JL, Peng ZM, Yang SC, Wagner FT, Yang H (2010) Truncated octahedral Pt(3)Ni oxygen reduction reaction electrocatalysts. J Am Chem Soc 132(14):4984–4985
9. Cailuo N, Oduro W, Kong ATS, Clifton L, Yu KMK, Thiebaut B, Cookson J, Bishop P, Tsang SC (2008) Engineering preformed cobalt-doped platinum nanocatalysts for ultraselective hydrogenation. ACS Nano 2(12):2547–2553
10. Wu YE, Wang DS, Zhao P, Niu ZQ, Peng Q, Li YD (2011) Monodispersed Pd-Ni nanoparticles: composition control synthesis and catalytic properties in the Miyaura-Suzuki reaction. Inorg Chem 50(6):2046–2048
11. Ueno S, Shimizu R, Kuwano R (2009) Nickel-catalyzed formation of a carbon-nitrogen bond at the beta position of saturated ketones. Angew Chem Int Edit 48(25):4543–4545
12. Harris PJF (1986) Sulfur-induced faceting of platinum catalyst particles. Nature 323(6091):792–794
13. Xu R, Xie T, Zhao Y, Li Y (2007) Quasi-homogeneous catalytic hydrogenation over monodisperse nickel and cobalt nanoparticles. Nanotechnology 18:055602
14. Schatz A, Reiser O, Stark WJ (2010) Nanoparticles as semi-heterogeneous catalyst supports. Chem-Eur J 16(30):8950–8967
15. Serna P, Corma A (2006) Chemoselective hydrogenation of nitro compounds with supported gold catalysts. Science 313(5785):332–334

16. Zhu Y, Qian HF, Drake BA, Jin RC (2010) Atomically precise $Au_{25}(SR)_{18}$ nanoparticles as catalysts for the selective hydrogenation of alpha, beta-unsaturated ketones and aldehydes. Angew Chem Int Edit 49(7):1295–1298

17. Wu Y, Cai S, Wang D, He W, Li Y (2012) Syntheses of water-soluble octahedral, truncated octahedral, and cubic Pt–Ni nanocrystals and their structure-activity study in model hydrogenation reactions. J Am Chem Soc 134(21):8975–8981

18. Tan Y, Xue X, Peng Q, Zhao H, Wang T, Li Y (2007) Controllable fabrication and electrical performance of single crystalline Cu_2O nanowires with high aspect ratios. Nano Lett 7(12):3723–3728

19. Tsung C-K, Kuhn JN, Huang W, Aliaga C, Hung L-I, Somorjai GA, Yang P (2009) Sub-10 nm platinum nanocrystals with size and shape control: catalytic study for ethylene and pyrrole hydrogenation. J Am Chem Soc 131(16):5816–5822

20. Bratlie KM, Lee H, Komvopoulos K, Yang P, Somorjai GA (2007) Platinum nanoparticle shape effects on benzene hydrogenation selectivity. Nano Lett 7(10):3097–3101

21. Stoner EC, Wohlfarth EP (1991) A mechanism of magnetic hysteresis in heterogeneous alloys. IEEE T Magn 27(4):3475–3518 (Reprinted from Philosophical Transaction Royal Society-London, vol 240, pp 599–642, 1948)

22. Ku JY, Aruguete DM, Alivisatos AP, Geissler PL (2011) Self-assembly of magnetic nanoparticles in evaporating solution. J Am Chem Soc 133(4):838–848

23. Alphandéry E, Ding Y, Ngo AT, Wang ZL, Wu LF, Pileni MP (2009) Assemblies of aligned magnetotactic bacteria and extracted magnetosomes: what is the main factor responsible for the magnetic anisotropy? ACS Nano 3(6):1539–1547

24. Petit C, Russier V, Pileni MP (2003) Effect of the structure of cobalt nanocrystal organization on the collective magnetic properties. J Phys Chem B 107(38):10333–10336

25. Wang DS, Li YD (2010) One-pot protocol for Au-based hybrid magnetic nanostructures via a noble-metal-induced reduction process. J Am Chem Soc 132(18):6280–6281

26. Ahrenstorf K, Albrecht O, Heller H, Kornowski A, Gorlitz D, Weller H (2007) Colloidal synthesis of Ni_xPt_{1-x} nanoparticles with tuneable composition and size. Small 3(2):271–274

27. Hou YL, Wang C, Kim JM, Sun SH (2007) A general strategy for synthesizing FePt nanowires and nanorods. Angew Chem Int Edit 46(33):6333–6335

28. Zhang J, Fang JY (2009) A general strategy for preparation of Pt 3d-transition metal (Co, Fe, Ni) nanocubes. J Am Chem Soc 131(51):18543–18547

29. Zhang J, Yang HZ, Fang JY, Zou SZ (2010) Synthesis and oxygen reduction activity of shape-controlled Pt(3)Ni nanopolyhedra. Nano Lett 10(2):638–644

30. Wu JB, Gross A, Yang H (2011) Shape and composition-controlled platinum alloy nanocrystals using carbon monoxide as reducing agent. Nano Lett 11(2):798–802

31. Wu BH, Zheng NF, Fu G (2011) Small molecules control the formation of Pt nanocrystals: a key role of carbon monoxide in the synthesis of Pt nanocubes. Chem Commun 47(3):1039–1041

32. Katoh K, Schmid GM (1971) Adsorption of benzoic acid on gold in perchlorate solutions. B Chem Soc Jpn 44(8):2007

33. Horanyi G, Solt J, Nagy F (1971) Investigation of adsorption phenomena on platinized Pt electrodes by tracer methods. 4. Adsorption of benzoic acid, benzenesulfonic acid, and phenylacetic acid. Acta Chim Hung 67(4):425

34. Neuber M, Zharnikov M, Walz J, Grunze M (1999) The adsorption geometry of benzoic acid on Ni(110). Surf Rev Lett 6(1):53–75

35. Pang X-Y, Lin R-N (2010) Adsorption mechanism of expanded graphite for oil and phenyl organic molecules. Asian J Chem 22(6):4469–4476

36. Koh M, Nakajima T (2000) Adsorption of aromatic compounds on C_xN-coated activated carbon. Carbon 38(14):1947–1954

37. Mazumder V, Sun SH (2009) Oleylamine-mediated synthesis of Pd nanoparticles for catalytic formic acid oxidation. J Am Chem Soc. 131 (13):4588–4589

38. Mostafa S, Behafarid F, Croy JR, Ono LK, Li L, Yang JC, Frenkel AI, Cuenya BR (2010) Shape-dependent catalytic properties of Pt nanoparticles. J Am Chem Soc 132(44):15714–15719
39. Tsunoyama H, Ichikuni N, Sakurai H, Tsukuda T (2009) Effect of electronic structures of Au clusters stabilized by Poly(N-vinyl-2-pyrrolidone) on aerobic oxidation catalysis. J Am Chem Soc 131(20):7086–7093
40. Magdesieva TV, Nikitin OM, Levitsky OA, Zinovyeva VA, Bezverkhyy I, Zolotukhina EV, Vorotyntsev MA (2012) Polypyrrole–palladium nanoparticles composite as efficient catalyst for Suzuki-Miyaura coupling. J Mol Catal A: Chem 353–354:50–57

Chapter 3
A Strategy for Designing a Concave Pt–Ni Alloy Through Controlled Chemical Synthesis

3.1 Introduction

During the past few decades, substantial advances in the so-called "bottom-up" synthesis has been achieved to precisely manipulate the atoms, which has allowed lots of preparations of semiconductor [1, 2] or metallic [3–5] nanoparticles that have well-defined structures and fascinating properties. Moreover, with stepwise bottom-up strategy, it is possible to open up the synthetic route to produce core–shell [6, 7], branched [8], alloy [9], and hybrid [10]. Combining the "bottom-up" strategy and a subsequent "top-down" carving process, a class of Au–Ag hollow and framework structures has been synthesized through the galvanic replacement method [11]. Based on this train of thought, polymetallic hollow NPs with higher complexity could be produced by a sequential galvanic exchange and a Kirkendall growth method [12]. These metal nanostructures with characteristic high surface area and hollow interior have achieved great success in optical sensing [13], catalysis [14], and drug delivery [15].

Through selectively etching away the less-noble metal and the subsequent rearrangement of the remaining metal atoms, the chemical etching approach has been shown to be an effective and classical "top-down" strategy of synthesis, which can simultaneously control the shape, size, and composition of metallic nanostructures [16, 17]. In 1926, the industrial catalyst Raney nickel was developed by selectively removing most of the aluminum from an Ni–Al alloy, while the active components were the remaining Ni atoms. This "old" material with a porous structure has been used as an important catalyst for the hydrogenation of vegetable oils to this day [18]. Inspired by this approach, our group synthesized a nanoporous Pt–Ni alloy recently, achieving an enhanced catalytic activity [19]. However, chemical etching suffers from some intrinsic drawbacks, including that the surface atoms of alloy are usually etched in random sites [20] and the etching process is usually too drastic to control, for which we could merely obtain porous structure without

© Springer-Verlag Berlin Heidelberg 2016
Y. Wu, *Controlled Synthesis of Pt–Ni Bimetallic Catalysts
and Study of Their Catalytic Properties*, Springer Theses,
DOI 10.1007/978-3-662-49847-7_3

precise control [21, 22]. Herein, it is still a great challenge for nanotechnology to make the chemical etching technique moderate and controllable. We predict that, there must be some unexpected structures and even certain breakthroughs in the field of nano-catalysis if we could rationally manipulate both the "bottom-up" and "top-down" processes to control the nanostructure.

In this chapter, we used coordinating complexes to control the chemical etching process at room temperature to synthesize a concave structure of Pt–Ni alloy. To further understand the nature of the controllable etching process, we verified our assumptions by using density functional theory (DFT) to support experimental observations. We found that there were some differences among the atomic cohesive energy (Ecoh) of Ni on specific sites (such as corner, edge, and face) in the alloys, which was identified as the factor governing the different etching priorities during the etching process. Next, we could effectively control the chemical etching process and design the desired structure based on the theoretical investigations and experimental data. Remarkably, owing to larger surface area and higher density of exposed atomic steps and defects, the concave nanostructures exhibited superior activity for catalytic reactions.

3.2 Experimental Section

3.2.1 Chemicals and Instruments

Chemicals. Analytical grade benzyl alcohol, benzoic acid, dimethylglyoxime, and ethylic acid were obtained from Beijing Chemical Reagents, P.R. China. Pt(acac)$_2$ (99 %), Ni(acac)$_2$ (99 %), and PVP (MW = 8000, AR) were purchased from Alfa Aesar. Pt/C (10 wt%) was purchased from Sigma Aldrich, activated carbon (Vulcan XC-72) was purchased from Carbot. All of the chemicals were used without further purification.

Instruments. XRD patterns were recorded by Rigaku D/Max 2500Pc X-ray powder diffractometer with CuKa radiation ($\lambda = 1.5418$ Å). TEM images were recorded by a JEOL JEM-1200EX working at 100 kV. HRTEM images were recorded by a FEI Tecnai G2 F20 S-Twin high-resolution transmission electron microscope working at 200 kV and a FEI Titan 80-300 transmission electron microscope equipped with a spherical aberration (Cs) corrector for the objective lens working at 300 kV. Electrochemical measurements were conducted on CH Instrument 660D electrochemical analyzer. X-ray photoelectron spectroscopy (XPS) experiments were performed on a ULVAC PHI Quantera microprobe. Binding energies (BE) were calibrated by setting the measured BE of C1s to 284.8 eV. Brunauer–Emmett–Teller (BET) surface area was measured with N$_2$ at 77 K by using a Quantachrome Autosorb-1 instrument. The catalytic reaction results were measured by gas chromatography (GC) (SP-6890) and gas chromatography–mass spectroscopy (GC–MS) (ITQ 700/900/1100) and 1H NMR. The NMR spectroscopy was conducted on a JEOL JNM-ECX 400 MHz instrument.

3.2.2 Experimental Methods

Preparation of Pt–Ni Octahedral Nanocrystals. In a typical synthesis of PtNi$_2$ octahedral nanocrystals, Pt(acac)$_2$, (8.0 mg), poly(vinylpyrrolidone) (PVP, MW = 8000), (80.0 mg), Ni(acac)$_2$ (10.0 mg), benzoic acid (50 mg), and benzylalcohol, (5 ml) were added to a 12-ml Teflon-lined stainless-steel autoclave, followed by 5–10 min vigorous stirring at room temperature. The sealed vessel was then heated at 150 °C for a 12 h. When it was cooled down to room temperature, the products were first precipitated by excess acetone, separated via centrifugation, and further purified by an ethanol-acetone mixture for three times.

Controllable Etching Process of Pt–Ni Octahedrons. The as-prepared Pt–Ni (4 mg) alloy was dispersed in H$_2$O (1 mL) and dimethylglyoxime (10 mg dissolved in 1 mL ethanol) added. The reaction mixture was stirred for 12 h. Acetic acid (5 mL; 50 %) was added and stirred for a further 15 min. The products were collected by centrifugation and further washed by ethanol for three times.

Electrochemical measurement. In a 100-ml beaker, the conductive activated carbon (20 mg) was dispersed in ethanol (20 ml) together with the as-prepared Pt–Ni (2 mg) alloy, followed by sonication for 30 min to ensure the catalysts were sufficiently incorporated onto activated carbon and stirring overnight at room temperature. The black mixture was separated via centrifugation with the dumping of the supernatant and the solid obtained was further purified by excess deionized water (30 ml) for three times. After drying and weighing, the sample was dispersed in water and diluted to 2 mg/ml.

Polishing electrode: The cleaned electrode was polished by gently drawing the number eight at a constant rate on the cloth with a small quantity of wetted Al$_2$O$_3$ powder, followed by washing the electrode with distilled water, being sonicated for 2 s in distilled water, ethanol, and washing the electrode with deionized water, sequentially. Next, cyclic voltammetry (CV) measurements were carried out in 0.001 M KFe(CN)$_6$ and 0.1 M KNO$_3$ with a scan rate 50 mv/s and a potential range between 0.60 and −0.20 V. However, it is necessary to repeat the procedures above if the potential difference of reversible peak was higher than 80 mV, until achieving stabilization around 80 mV. Finally, drying up the Glassy Carbon Electrode(GCE) after cleaning up the electrode.

Sample Application: The sample was sonicated for 5 min. Then, 5 ul of the dispersion was transferred onto the disk of the GCE. After the working electrode was dried, 5 μl Nafion dilutes (0.05 wt%) was coated on the catalyst' surface. Electrolyte (50 ml, 0.1 M HClO$_4$ or 0.5 M H$_2$SO$_4$ solutions) was transferred into a sealed vessel (such as centrifuge tube or conical flask with plug) which was firmly fixed. Then, in a typical procedure, a hosepipe with a 5-ml pipette was linked with a nitrogen cylinder; then, the solution was purged with N$_2$ for 30 min by inserting the pipette into the solution, with stable bubbles observed.

The working electrode was a Glassy Carbon Electrode (GCE) (dimeter: 5 mm and area: 0.196 cm^2) from Pine research Instrument. A leak-free AgCl/Ag/KCl (saturated) electrode was used as the reference and a Pt wire was used as the counter electrode. First, cyclic voltammetry (CV) measurements were carried out in 0.1 M HClO$_4$ with scan rate 100mv/s for 50 cycles to activate the system; then, the electrochemical active surface area was estimated by CV with a scan rate 0.05 v/s and a potential range between −0.2 and 1.0 v vs. RHE)

In contrast, 0.6 ml carbinol was added to the as-prepared electrolyte (0.1 M HClO$_4$ or 0.5 M H$_2$SO$_4$ solution) with mild stirring; then, CV measurements were carried out with a scan rate 50 mv/s for 15 cycles or another 20–30 cycles until the peak current verged to constant.

The hydrogenation of nitrobenzene. First, 51 μL nitrobenzene (0.5 mmol) in THF (2.5 mL) and the Pt–Ni alloy (contain 0.0025 mmol Pt, 0.5 mol%) were added in a 10-mL round flask. The round flask was purged with H$_2$; then, the reaction mixture was immersed in an oil bath at 25 °C. The product was subsequently characterized by GC and 1H NMR.

3.3 Results and Discussion

3.3.1 Structural Characterization and Discussion

Figure 3.1a showed transmission electron microscope (TEM) image of the star-like corroded PtNi$_2$ NPs which have six uniform arms. As shown in the magnified TEM image and the high-angle annular dark-field scanning transmission electron microscope (HAADF-STEM) micrograph (Fig. 3.1b–d), the six arms were much brighter than the center region of the particles and each face of NPs was excavated resulting in a curved cavity, indicating the presence of the concave structure. The corresponding elemental maps of HAADF-STEM micrograph showed that both Pt and Ni were distributed evenly throughout each individual concave nanocrystal.

The high-resolution TEM (HRTEM) images of a single concave Pt–Ni NP from different directions (Figs. 3.1c and 3.2) clearly showed the surface contained rich steps and defects. Previous synthesis have allowed us to prepare a series of octahedral Pt–Ni nanocrystals (average size: 11.8 nm) strictly bounded by six {111} facets [24].

The octahedral Pt–Ni NPs have excellent monodispersity in various polar solvents owing to the coating of the hydrophilic surfactant molecules, which benefits the interaction with the water-soluble ligand (dimethylglyoxime) in the subsequent chemical etching procedure.

Figure 3.3 shows TEM and HRTEM images of Pt–Ni NPs prepared with increasing concentrations of Ni (Ni:Pt mole ratio was varied from 2:1 to 3:1 and 10:1). We could observe that, since more and more Ni could be dissolved from Ni-rich Pt–Ni alloy, the concavity of the obtained nanocrystals significantly increased as the nickel content of the starting alloys increased. Powder X-ray diffraction (XRD) measurements of Pt–Ni NPs were used to identify the change of the internal crystalline structures during the chemical etching process and the Fig. 3.4 showed that the diffraction patterns of the Pt–Ni alloys could be indexed to {111}, {200}, and {220} diffractions of a face-centered-cubic (fcc) structure.

Fig. 3.1 **a** TEM image and **b** magnified TEM image of corroded PtNi₂. Inset is the ideal model of a concave octahedron. **c** HRTEM image of corroded PtNi₂ orientated along {111} direction. Insets are the FFT (fast Fourier transition) patterns and corresponding lattice ball models. **d** HAADF-STEM image and corresponding elemental maps (Pt and Ni) of corroded PtNi₂ particles. Reproduced from Ref. [23] by permission of John Wiley & Sons Ltd

After chemical etching, all the peaks of the corroded Pt–Ni alloys shifted slightly to a lower 2θ value compared to those of uncorroded Pt–Ni alloys, which could be attributed to the increased lattice spacing resulting from most of Ni being etched from the lattice of Pt–Ni alloys. Intriguingly, the XRD results suggested that all the initially Ni-rich Pt–Ni alloys were gradually eroded into concave structures, all with the composition Pt_3Ni, which indicated that the etching process came to a stop when the molar ratio of Pt/Ni was three. The accurate compositions were further verified by energy dispersive spectroscopy (EDS) and inductively coupled plasma-mass spectrometry (ICP-MS).

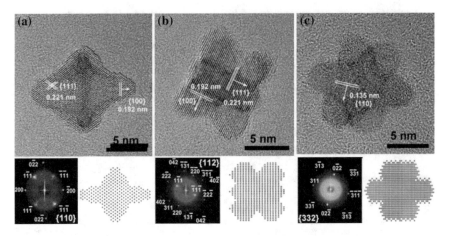

Fig. 3.2 HRTEM image of concave Pt–Ni NPs from different directions. Reproduced from Ref. [23] by permission of John Wiley & Sons Ltd

3.3.2 Etching Mechanism

In the synthesis of the concave Pt–Ni bimetallic nanostructure, the dimethylgly-oxime played a crucial role. The chemical etching process can be described by the chemical equilibrium Eqs. (3.1)–(3.3).

$$\text{Ni}\,(0) \; - \; 2e^- \;\Longleftrightarrow\; \text{Ni}\,(\text{II}) \tag{3.1}$$

$$1/2\,\text{O}_2 \; + \; \text{H}_2\text{O} \; + 2\,e^- \;\Longleftrightarrow\; 2\text{OH}^- \tag{3.2}$$

$$\text{Ni}\,(\text{II}) + 2 \;\; \underset{\text{C}}{\overset{\text{C}}{\big|}}\!\!\begin{matrix}\text{N–OH}\\ \text{N–OH}\end{matrix} \;\Longleftrightarrow\; \left[\begin{matrix} \text{OH} & \text{OH} \\ \text{N} & \text{N} \\ \text{C} & \text{Ni} & \text{C} \\ \text{N} & \text{N} \\ \text{OH} & \text{OH} \end{matrix} \right]^0 + 2\text{H}^+ \tag{3.3}$$

Equations (3.1) and (3.2) could be considered as two half-reactions of an oxidation-reduction reaction. Referring to the corresponding operating processes and experimental observations, the Ni atoms on the bimetallic surface were oxidized to Ni(II) by the oxygen in the air (Fig. 3.5).

To test this hypothesis, we subsequently detected the concentration of Ni ions during the etching process in situ. The results showed that the rate of the etching process could be enhanced by the increased concentration of oxygen. When oxygen was replaced with nitrogen, the etching process barely occurred, which corresponded to a process of oxidation etching (Fig. 3.6).

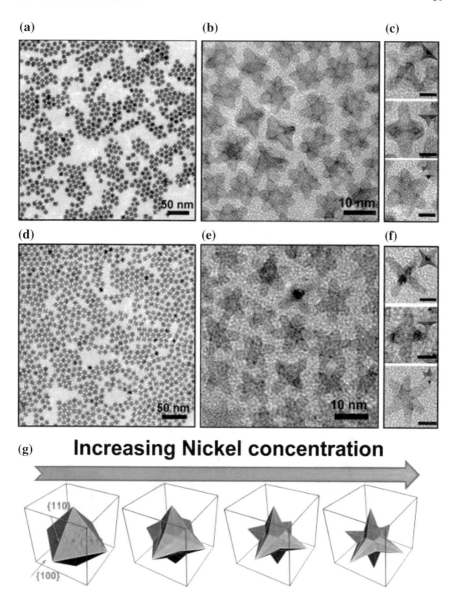

Fig. 3.3 TEM images of **a** corroded PtNi$_3$ and **d** corroded PtNi$_{10}$. Magnified TEM images of (**b**) corroded PtNi$_3$ and **e** corroded PtNi$_{10}$. **c**, **f** are the HRTEM images and corresponding models orientated along {100} (top panel), {110} (middle panel), and {111} (bottom panel) directions, respectively (scale bars are 5 nm). **g** The evolution of nanoparticle shape as a function of Ni:Pt mole ratio in the originally Ni-rich alloys. Reproduced from Ref. [23] by permission of John Wiley & Sons Ltd

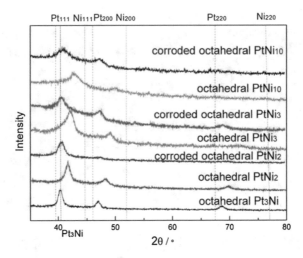

Fig. 3.4 XRD patterns of as-prepared Pt–Ni alloys and corroded Pt–Ni nanoparticles. Reproduced from Ref. [23] by permission of John Wiley & Sons Ltd

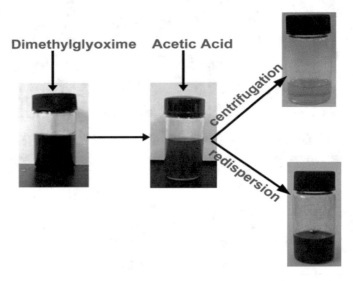

Fig. 3.5 Illustration of the typical procedure of the chemical etching. Reproduced from Ref. [23] by permission of John Wiley & Sons Ltd

For Eq. (3.3), the dimethylglyoxime selectively coordinates to Ni(II) rather than Pt(II) to generate a red dimethylglyoxime nickel precipitate under neutral conditions [25], indicating the higher susceptibility and dissolution rate of Ni species to be oxidized by oxygen compared to Pt. Quite strikingly, Pt species were not detected by ICP-MS measurements of the mother solution after washing and separation, implying Pt was not being etched from the staring polyhedrons through this

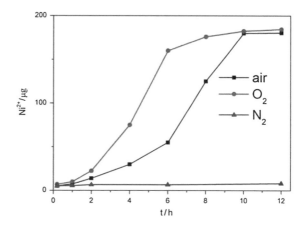

Fig. 3.6 Dependency of nickel concentration and reaction time in the presence of different gas. Reproduced from ref. [23] by permission of John Wiley & Sons Ltd

method. Furthermore, according to the Eq. (3.3), acetic acid could be added to dissolve the as-produced precipitate by shifting the chemical equilibrium to the left. The result could be traced and verified by the X-ray photoelectron spectroscopy (XPS) experiments. The surface information about octahedral PtNi$_3$ alloy revealed the Pt 4f7/2 and 4f5/2 had a binding energy of 71.3 and 74.6 eV, characteristic of Pt(0). The Ni 2p3/2 was 852.8 eV, which corresponded to the Ni(0). During the chemical etching process, the valence state of Pt is barely altered with respect to the initial data while there were oxidized Ni detected in the corroded PtNi$_3$ alloy. In contrast, the XPS results showed the complex were mostly washed by the addition of acetic acid (Fig. 3.7).

Surprisingly, the chemical etching process did not occur without assistance of dimethylglyoxime even at an elevated temperature (100 °C). However, the dimethylglyoxime could not completely etch all the Ni atoms from Pt–Ni alloys, even if the reaction time was lengthened or the dosage of dimethylglyoxime was increased. These interesting phenomena could be understood by carrying out measurements on redox potentials of Pt–Ni NP decorated electrode. The oxidation potentials (vs. Ag/AgCl) of a Pt–Ni NP decorated electrode decreased when the Ni concentration increased or dimethylglyoxime was introduced, indicating that the Pt–Ni alloy could be oxidized much easier, that was because the active component Ni could lose electrons and then dissolve from NPs much easier. As the chemical etching proceeded, the oxidation potential of the Ni-rich alloy increased with the increasing ratio of Pt to Ni, revealing that as the alloy corroded it became more difficult to be etched and finally terminated in Pt$_3$Ni (Fig. 3.8).

The controllability of this chemical etching method was shown to be essential for the formation of this metastable concave structure. On the one hand, we used octahedral PtNi$_{10}$ as a probe to gain insight into the morphological evolution during the etching process. Figure 3.9 shows a series of TEM images of NPs with an increasing concavity could be obtained during different stages of chemical etching.

As the chemical etching proceeded, a series of NPS with a ceaselessly increasing concavity could be obtained. However, in the initial stage of the etching

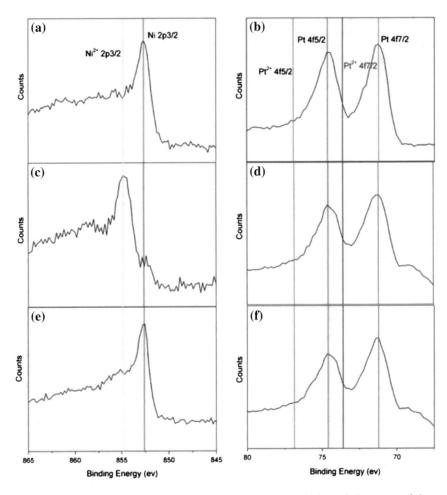

Fig. 3.7 XPS data for the initial octahedral PtNi₃ nanocrystals: high-resolution scans of the **a** Ni 2p and **b** Pt 4f. XPS data for the corroded octahedral PtNi₃ before washed with acetic acid: high-resolution scans of the **c** Ni 2p and **d** Pt 4f. XPS data for the corroded octahedral PtNi₃ after washed with acetic acid: high-resolution scans of the **e** Ni 2p and **f** Pt 4f. Reproduced from Ref. [23] by permission of John Wiley & Sons Ltd

process, the etching process prevailed along the {100} direction, which suggested the etching process started at the corners first. When the new narrow {100} facets formed, the etching shifted to the {110} and {111} directions, and the edges and the facets would be excavated to curved cavities with the generation of the concave structure. On the other hand, the severely corrosive conditions, such as concentrated nitric acid treatment, usually distort and collapse this concave structure with the result of some tangle structures (Fig. 3.10).

As we all know, chemical etching reflects, to a certain extent, the stability of the metal atoms of NPs in the reaction environment, which could be correctly

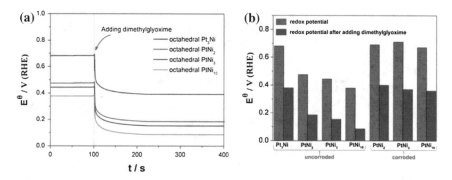

Fig. 3.8 The electronical potentials of the as-prepared Pt–Ni alloys decorated working electrodes. **a** time-dependent curve of potential analysis. **b** statistic potentials of as-prepared Pt–Ni all. Reproduced from Ref. [23] by permission of John Wiley & Sons Ltd

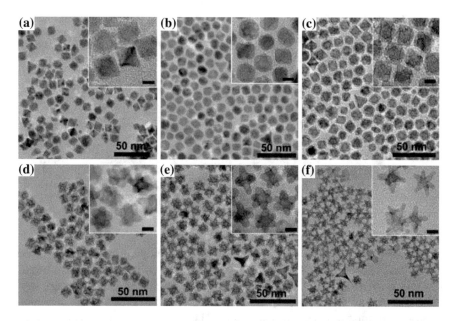

Fig. 3.9 Shape evolution of octahedral PtNi$_{10}$ at different chemical etching stage **a** 15 min, **b** 30 min, **c** 1 h, **d** 3 h, **e** 6 h, **f** 12 h. *Insets* are the corresponding HRTEM images (scale bars are 5 nm). Reproduced from Ref. [23] by permission of John Wiley & Sons Ltd

described by the Ecoh of alloys. As shown in Table 3.1, the value of Ecoh not only varied with the nature of metal (Ni or Pt) but also with the geometrical site [corner (C), edge (E), or face (F)].

From the thermodynamic point of view, the calculated Ecoh showed the possibility of removal of metal atoms from the surface of alloys decreased in the order of CNi > ENi(Ni) > ENi(Ni–Pt) > CPt > FNi > EPt(Ni–Pt) > FPt in the same

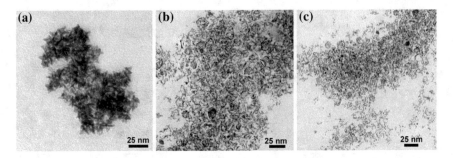

Fig. 3.10 TEM images of **a** corroded octahedral PtNi$_2$, **b** corroded octahedral PtNi$_3$ nanocrystals, and **c** corroded octahedral PtNi$_{10}$ prepared by replacing dimethylglyoxime with excess amount of concentrated nitric acid. Reproduced from Ref. [23] by permission of John Wiley & Sons Ltd

Table 3.1 Calculated atomic cohesion energies (in eV/per atom) by density functional theory, and the coordination number and coordinated atoms from crystallographic analysis of Ni and Pt atoms at the corner (C), edge (E) and face (F) of octahedral PtNi$_3$ nanoparticles as compared to {100}, {110} and {111} surfaces. Subscripts (Ni) and (PtNi) mean 100 % Ni and 50 % Ni + 50 % Pt surface (face or edge), respectively

	Site	Ecoh	Pt–Ni	Ni–Ni	Surface	Site	Ecoh	Pt–Ni	Ni–Ni
Octahedral PtNi$_3$ NPs	CNi	3.39	2	2	(100)	SNi (Ni)	4.64	2	6
	CPt	4.94	4			S Ni (PtNi)	4.91	4	4
	ENi (Ni)	4.66	2	5		SPt (PtNi)	6.03	8	
	ENi (PtNi)	4.69	3	4	(110)	SNi (Ni)	4.55	2	5
	EPt (PtNi)	6.28	7			S Ni (PtNi)	4.52	3	4
	FNi	5.34	3	6	(111)	SNi	5.24	3	6
	FPt	6.69	9			SPt	6.81	9	

reaction environment. The schematic illustration of the proposed etching process of PtNi$_3$ NPs based on DFT calculations and structural dynamics analysis are summarized in Fig. 3.11. We think that the Ni atoms on the corners and edges were preferably removed, followed by the deposition and segregation of the remaining Pt atoms, forming round corners and steps. This result was in agreement with the HRTEM observations in Fig. 3.9. While for the {111} surface of the octahedron, the Ni atoms were present in much greater numbers than the Pt atoms, so the dissolution rate of the Ni atoms was much more pronounced than the diffusion-segregation rate of the Pt atoms, indicating the predominant role of exfoliation etching on the {111} surface.

Furthermore, for the same kind of atoms, the faces are more stable sites than the corners and edges because of the Ecoh of the faces is higher than those of the corners and edges. Thus, the etching started from corner sites rather than {111} faces. The proposed etching process and the increasing Pt/Ni ratio of corroded NPs were in agreement with the experimental observations. Noted that only

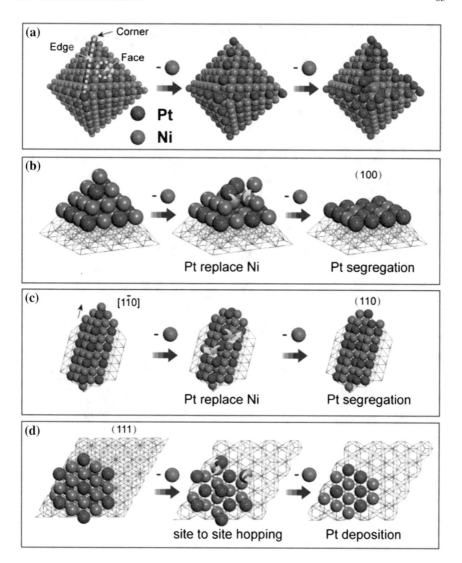

Fig. 3.11 Schematic illustration of the proposed etching mechanism of PtNi₃ octahedron. Reproduced from Ref. [23] by permission of John Wiley & Sons Ltd

Pt-segregation occurs in corroded NPs because the Ni atoms were dissolved in the solution, which was quite different from the cases of Pt₃Ni single crystals [26]. DFT calculations also showed that the density of states (DOS) at (and near) the Fermi level of the corroded surfaces was more than twice (and four times) those of {111} surface of Pt₃Ni single-crystal, indicating that concave Pt₃Ni NPs had a higher activity than single-crystal ones. Another type of Pt–Ni alloy with a predictable concave cubic morphology was also obtained by our method, if the original octahedron was replaced with a PtNi₃ cube (Fig. 3.13).

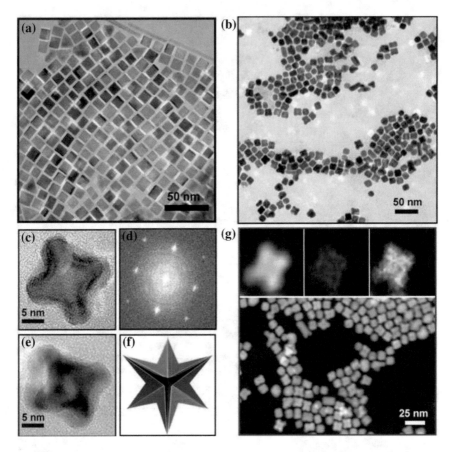

Fig. 3.12 **a** TEM image of cubic PtNi3, **b** large-area TEM, **c** HRTEM, and **d** tilted HRTTEM images of the as-prepared concave cubic Pt3Ni. Corresponding **d** FFT pattern and **e** model of concave cubic Pt3Ni. **f** HAADF-STEM image and corresponding **d** elemental maps (Pt (*yellow*) and Ni (*red*)) of concave cubic Pt3Ni. Figure 3.12 **a** TEM image of cubic PtNi3, **b** large-area TEM, **c** HRTEM, and **d** tilted HRTTEM images of the as-prepared concave cubic Pt3Ni. Corresponding **d** FFT pattern and **e** model of concave cubic Pt3Ni. **f** HAADF-STEM image and corresponding **d** elemental maps (Pt (*yellow*) and Ni (*red*)) of concave cubic Pt3Ni. Reproduced from Ref. [23] by permission of John Wiley & Sons Ltd

3.3.3 The Catalytic Properties of Concave Pt–Ni Polyhedron

Brunauer–Emmett–Teller (BET) measurements were used to analyze the surface area of octahedral Pt–Ni NPs and corroded Pt–Ni alloys. The chemical etching treatment gave the corroded PtNi3 (64 m^2 g^{-1}) a larger surface area than the octahedral PtNi3 (38 m^2 g^{-1}). To verify the higher electrocatalytic active of the concave structure, the cyclic voltammetry (CV) and electrocatalytic oxidation of methanol were used to evaluate the electrochemical activity of as-synthesized

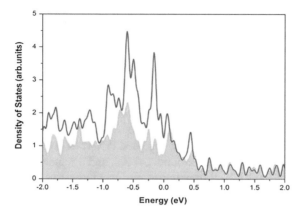

Fig. 3.13 The density of states of {111} surface of single-crystal NPs (*green filled areas*) and corroded surface of concave Pt$_3$Ni NPs (*red line*) by DFT calculations. The Fermi level was set to zero. Reproduced from Ref. [23] by permission of John Wiley & Sons Ltd

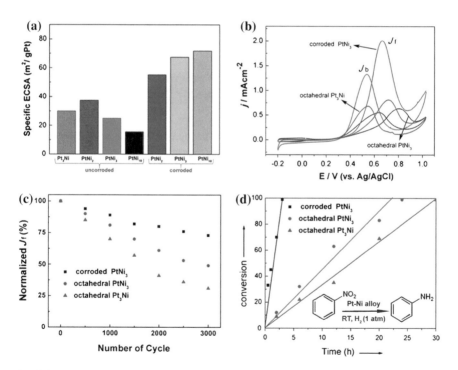

Fig. 3.14 a Specific ECSAs for Pt–Ni alloys recorded in 0.1 m HClO$_4$ and **b** cyclic voltammograms of methanol oxidation on Pt–Ni octahedrons and corroded Pt–Ni alloys in 0.1 m HClO$_4$ containing 1 m MeOH (f = forward scan, b = backward scan). **c** Loss of peak current density in a forward scan as a function of cycling numbers. **d** Conversion % as a function of time in hydrogenation of nitrobenzene catalyzed by octahedral PtNi$_3$, octahedral Pt$_3$Ni, and corroded PtNi$_3$ nanocrystals. Reproduced from Ref. [23] by permission of John Wiley & Sons Ltd

Pt–Ni alloys. The corroded PtNi$_{10}$ showed the highest specific electrochemically active surface area (ECSA) (71.5 m^2 gPt$-$1) by measuring the charge collected in the electrochemical absorption/desorption region of 0–0.40 V and assuming a value of 210 uC cm^{-2} for the adsorption of a monolayer of hydrogen on platinum. No matter it was the octahedral PtNi$_3$ or the octahedral PtNi$_2$, there were a certain enlargement of ESCA of the corroded polyhedron, which was 67.5 m^2 gPt$-$1 and 52.3 m^2 gPt$-$1, respectively, results which correlate with the degree of concavity. Figure 3.14 showed that the peak current densities of methanol oxidation in the forward (positive) potential (Jf) were: corroded PtNi$_3$ > corroded PtNi$_2$ > corroded PtNi$_{10}$ which were, respectively, around 2.4, 3.2, and 3.8 times higher than for the uncorroded octahedral NPs. Furthermore, the specific activity of corroded PtNi$_3$ was still at 73 % of the original value after the durable test with 3000 cycle, which was higher than for the durable test of the octahedral PtNi$_3$ and Pt$_3$Ni. Methanol can be oxidized to CO species which poison the Pt catalyst sites and limit their activity [27]. Considering atomic steps on Pt offer a superior activity for the oxidation of CO [28], thus the improved activity of methanol oxidation on concave Pt–Ni alloys could be attributed to the high density of Pt-segregated atomic steps. To further confirm that the corroded Pt–Ni alloys have a better tolerance to poisoning CO, CO stripping experiments were conducted, which had verified the enhanced CO-removal ability of the concave Pt–Ni alloys by the oxidation of CO.

The structure of the concave Pt–Ni alloy also suggested potential as a heterogeneous catalyst owing to its specificity. We chose the corroded PtNi$_3$, octahedral Pt$_3$Ni, and PtNi$_3$ as probes to study the structure-activity dependences

Table 3.2 Turnover frequencies (TOFs [$h - 1$]) of the hydrogenation of nitroarenes catalyzed by various catalysts

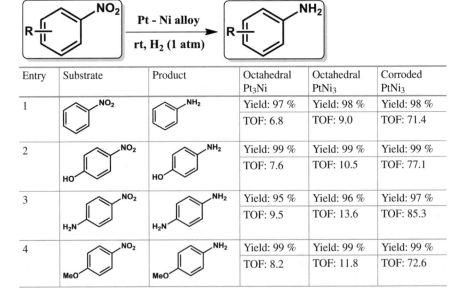

Entry	Substrate	Product	Octahedral Pt$_3$Ni	Octahedral PtNi$_3$	Corroded PtNi$_3$
1			Yield: 97 %	Yield: 98 %	Yield: 98 %
			TOF: 6.8	TOF: 9.0	TOF: 71.4
2			Yield: 99 %	Yield: 99 %	Yield: 99 %
			TOF: 7.6	TOF: 10.5	TOF: 77.1
3			Yield: 95 %	Yield: 96 %	Yield: 97 %
			TOF: 9.5	TOF: 13.6	TOF: 85.3
4			Yield: 99 %	Yield: 99 %	Yield: 99 %
			TOF: 8.2	TOF: 11.8	TOF: 72.6

in the hydrogenation reactions of nitroarenes, which are important industrial process. Table 3.2 showed that the nitrobenzene [29] was completely converted into aniline with nearly 100 % selectivity in 3 h with the catalysis by corroded PtNi$_3$. However, the calculated initial activities turnover frequency (TOF) of corroded PtNi$_3$ was 7.9, and 10.5-times higher than that catalyzed by octahedral PtNi$_3$ and Pt$_3$Ni. Herein, the Pt–Ni catalysts with concave structure might play a crucial role in various organic reactions just as an excellent heterogeneous catalyst.

3.4 Conclusions

In summary, this work has presented a controllable "top-down" synthesis of concave Pt–Ni alloys through a coordination-assisted chemical etching process. Based on the experimental observations and DFT calculations, the formation of the concave structure was attributed to the different etching priorities on specific sites. Owing to the larger surface area and higher density of exposed atomic steps, concave Pt–Ni alloys exhibited higher catalytic activity for the electrocatalytic oxidation of methanol and the hydrogenation of nitroarene compared to the uncorroded octahedral Pt–Ni alloys. The development of this synthetic strategy would provide effective tool for us to design much more efficient and practical catalysts in the near future.

References

1. Murray C, Norris D, Bawendi MG (1993) Synthesis and characterization of nearly monodisperse CdE (E = sulfur, selenium, tellurium) semiconductor nanocrystallites. J Am Chem Soc 115(19):8706–8715
2. Peng X, Manna L, Yang W, Wickham J, Scher E, Kadavanich A, Alivisatos A (2000) Shape control of CdSe nanocrystals. Nature 404(6773):59–61
3. Sun S, Murray C, Weller D, Folks L, Moser A (2000) Monodisperse FePt nanoparticles and ferromagnetic FePt nanocrystal superlattices. Science 287(5460):1989–1992
4. Wang X, Zhuang J, Peng Q, Li Y (2005) A general strategy for nanocrystal synthesis. Nature 437(7055):121–124
5. Sun Y, Xia Y (2002) Shape-controlled synthesis of gold and silver nanoparticles. Science 298(5601):2176–2179
6. Habas SE, Lee H, Radmilovic V, Somorjai GA, Yang P (2007) Shaping binary metal nanocrystals through epitaxial seeded growth. Nat Mater 6(9):692–697
7. Ghosh Chaudhuri R, Paria S (2011) Core/shell nanoparticles: classes, properties, synthesis mechanisms, characterization, and applications. Chem Rev 112(4):2373–2433
8. Lim B, Jiang M, Camargo PHC, Cho EC, Tao J, Lu X, Zhu Y, Xia Y (2009) Pd–Pt bimetallic nanodendrites with high activity for oxygen reduction. Science 324(5932):1302–1305
9. Sra AK, Schaak RE (2004) Synthesis of atomically ordered AuCu and AuCu3 nanocrystals from bimetallic nanoparticle precursors. J Am Chem Soc 126(21):6667–6672
10. Macdonald JE, Sadan MB, Houben L, Popov I, Banin U (2010) Hybrid nanoscale inorganic cages. Nat Mater 9(10):810–815

11. Sun Y, Mayers BT, Xia Y (2002) Template-engaged replacement reaction: a one-step approach to the large-scale synthesis of metal nanostructures with hollow interiors. Nano Lett 2(5):481–485

12. González E, Arbiol J, Puntes VF (2011) Carving at the nanoscale: sequential galvanic exchange and Kirkendall growth at room temperature. Science 334(6061):1377–1380

13. Mulvihill MJ, Ling XY, Henzie J, Yang P (2009) Anisotropic etching of silver nanoparticles for plasmonic structures capable of single-particle SERS. J Am Chem Soc 132(1):268–274

14. Biener J, Wittstock A, Zepeda-Ruiz L, Biener M, Zielasek V, Kramer D, Viswanath R, Weissmüller J, Bäumer M, Hamza A (2008) Surface-chemistry-driven actuation in nanoporous gold. Nat Mater 8(1):47–51

15. Yavuz MS, Cheng Y, Chen J, Cobley CM, Zhang Q, Rycenga M, Xie J, Kim C, Song KH, Schwartz AG (2009) Gold nanocages covered by smart polymers for controlled release with near-infrared light. Nat Mater 8(12):935–939

16. Wittstock A, Zielasek V, Biener J, Friend C, Bäumer M (2010) Nanoporous gold catalysts for selective gas-phase oxidative coupling of methanol at low temperature. Science 327(5963):319–322

17. Mulvihill MJ, Ling XY, Henzie J, Yang P (2010) Anisotropic etching of silver nanoparticles for plasmonic structures capable of single-particle SERS. J Am Chem Soc 132(1):268–274

18. Raney M (1926) U.S. Patent

19. Wang D, Zhao P, Li Y (2011) General preparation for Pt-based alloy nanoporous nanoparticles as potential nanocatalysts. Sci Rep-Uk 1 47(1):1–5

20. Blonder G (1986) Simple model for etching. Phys Rev B 33(9):6157

21. Sieradzki K, Erlebacher J, Karma A, Dimitrov N, Aziz M (2001) Evolution of nanoporosity in dealloying. Nature 410(6827):450–453

22. Shui JI, Chen C, Li JCM (2011) Evolution of nanoporous Pt-Fe alloy nanowires by dealloying and their catalytic property for oxygen reduction reaction. Adv Funct Mater 21(17): 3357–3362

23. Wu Y, Wang D, Niu Z, Chen P, Zhou G, Li Y (2012) A strategy for designing a concave Pt–Ni alloy through controllable chemical etching. Angew Chem Int Ed 51(50):12524–12528

24. Wu J, Qi L, You H, Gross AM, Li J, Yang H (2012) Icosahedral platinum alloy nanocrystals with enhanced electrocatalytic activities. J Am Chem Soc 134(29):11880–11883

25. Gazda DB, Fritz JS, Porter MD (2004) Determination of nickel(II) as the nickel dimethylglyoxime complex using colorimetric solid phase extraction. Anal Chim Acta 508(1):53–59

26. Stamenkovic VR, Fowler B, Mun BS, Wang G, Ross PN, Lucas CA, Marković NM (2007) Improved oxygen reduction activity on Pt₃Ni(111) via increased surface site availability. Science 315(5811):493–497

27. Wasmus S, Küver A (1999) Methanol oxidation and direct methanol fuel cells: a selective review. J Electroanal Chem 461(1):14–31

28. Chen QS, Vidal-Iglesias FJ, Solla-Gullón J, Sun SG, Feliu JM (2011) Role of surface defect sites: from Pt model surfaces to shape-controlled nanoparticles. Chem Sci 3(1):136–147

29. Blaser HU, Steiner H, Studer M (2009) Selective catalytic hydrogenation of functionalized nitroarenes: an update. Chemcatchem 1(2):210–221

Chapter 4
Defect-Dominated Shape Recovery of Nanocrystals: A New Strategy for Trimetallic Catalysts

4.1 Introduction

Noble metals are of immense importance in diverse areas such as chemical industry [1], new energy resources [2], and gas sensors [3]. The properties of noble metals for use in heterogeneous catalysis are strongly dependent on their surface atomic structure [4]. The extensive investigations of defects on the noble metal surface (e.g., steps, kinks, and edges) have largely drawn public attention resulting from their predominant role in actual catalytic processes [5, 6]. These open structures, which have a high density of low-coordinate atoms, usually favor reducing the activation energy for structure-sensitive reactions [7]. For example, some typical reports have pointed out that the unsaturated coordinated surface atoms have the ability to give priority to the cleavage or formation of molecular π bonds [8]. Notably, recent progress in computational catalysis, surface science, and nanotechnology has significantly shown that the catalytic characteristics are dependent not only on geometrical effects but also on electronic effects [9].

For a hybrid structure such as metal–oxide or metal–metal, the interface perimeter has generally been viewed as the most active site for some model reactions such as low-temperature CO oxidation [10], water–gas shift reaction [11], etc. Furthermore, the formation of ternary metallic nanocrystals by incorporating a third metal into bimetallic structures has been shown to affect the adsorption of micromolecules and the transfer of electrons, which favors improving the catalytic property of catalysts [12]. Encouraged by this progress, it is promising to design multicenter and multifuncational catalysts to optimize the catalytic performance. Considering the different rates of nucleation and growth of different metals, traditional synthetic methodologies such as co-reduction, coprecipitation, and codecomposition are becoming insufficient to completely control the size, composition, structures, and synthesis of trimetallic catalysts [13]. Compared to the other surface sites of the nanoparticles, the defect sites are more susceptible to be

© Springer-Verlag Berlin Heidelberg 2016
Y. Wu, *Controlled Synthesis of Pt–Ni Bimetallic Catalysts and Study of Their Catalytic Properties*, Springer Theses, DOI 10.1007/978-3-662-49847-7_4

attacked by active species such as etchants, oxidants, and newly formed nuclei due to the lower interface energy [14]. Herein, rational use of the volatility of defects in crystal growth allows the third metal to be introduced and site-selectively anchored around the defects on the bimetallic surface, favoring controlling the geometrical and electronic structure of the trimetallic structures. However, surface defects, which usually serve as the most active sites for catalysis, normally vanish faster during the growth stage and thus are preferentially eliminated on the final surface owing to the lower surface energy [15]. Therefore, in the context of defect survival, precise manipulation of the growth of the third metal at the atomic level to achieve the optimal tradeoff of geometrical and electronic effects is very challenging.

4.2 Experimental Section

4.2.1 Chemicals and Instruments

Chemicals: Analytical grade benzyl alcohol, benzoic acid, dimethylglyoxime, chloroauric acid ($HAuCl_4$ $4H_2O$), silver nitrate ($AgNO_3$), and acetic acid were obtained from Beijing Chemical Reagents, China. platinum bis(acetylacetonate) ($Pt(acac)_2$), $Ni(acac)_2$, $Cu(acac)_2$, $Rh(acac)_3$, $Fe(acac)_3$, and poly (vinylpyrrolidone) (PVP, MW = 8000, AR) were purchased from Alfa Aesar. All of the chemicals used in this experiment were analytical grade and used without further purification.

Instruments: XRD patterns were recorded by Rigaku D/Max 2500Pc X-ray powder diffractometer with CuKa radiation ($\lambda = 1.5418$ Å). TEM images were recorded by a JEOL JEM-1200EX working at 100 kV. HRTEM images were recorded by a FEI Tecnai G2 F20 S-Twin high-resolution transmission electron microscope working at 200 kV and a FEI Titan 80–300 transmission electron microscope equipped with a spherical aberration (Cs) corrector for the objective lens working at 300 kV. Electrochemical measurements were conducted on CH Instrument 660D electrochemical analyzer. X-ray photoelectron spectroscopy (XPS) experiments were performed on a ULVAC PHI Quantera microprobe. Binding energies (BE) were calibrated by setting the measured BE of C1 s to 284.8 eV. The catalytic reaction results were measured by gas chromatography (GC) (SP-6890) and gas chromatography–mass spectroscopy (GC–MS) (ITQ 700/900/1100) and 1H NMR. The NMR spectroscopy was conducted on a JEOL JNM-ECX 400 MHz instrument.

4.2.2 Experiment Methods

Preparation of octahedral $PtNi_3$ nanocrystals. In a typical synthesis of $PtNi_3$ octahedral nanocrystals, $Pt(acac)_2$, (8.0 mg), poly(vinylpyrrolidone) (PVP, MW = 8000), (80.0 mg), $Ni(acac)_2$ (15.0 mg) and benzoic acid (50 mg), and benzylalcohol, (5 ml) were added to a 12-ml Teflon-lined stainless-steel autoclave, followed by 5–10 min vigorous stirring at room temperature. The sealed vessel was then heated at 150 °C for a 12 h. When it was cooled down to room temperature, the products were first precipitated by excess acetone, separated via centrifugation, and further purified by an ethanol–acetone mixture for three times.

Controllable Etching Process of Pt–Ni Octahedrons. The as-prepared Pt–Ni (4 mg) alloy was dispersed in H_2O (1 mL) and dimethylglyoxime (10 mg dissolved in 1 mL ethanol) added. The reaction mixture was stirred for 12 h. Acetic acid (5 mL; 50 %) was added and stirred for a further 15 min. The products were collected by centrifugation and further washed by ethanol for three times.

Second growth of Ni using concave Pt$_3$Ni as seeds. The 4 mg sample of concave Pt$_3$Ni alloy seeds was diluted by 5 mL benzoic alcohol to 0.8 mg/mL (containing 0.0185 mmol of Pt and 0.0064 mmol Ni base on ICP-MS measurement) and sonicated for 5 min. Then, 13.3 mg Ni(acac)$_2$ (0.052 mmol), 0.05 g benzoic acid, and 0.05 g PVP were added and further stirred for 10 min. The as-obtained turbid liquid was transferred into a 12-mL Teflon-lined stainless-steel autoclave. The sealed vessel was then heated at 150 °C for a 12 h. When it was cooled down to room temperature, the products were first precipitated by excess acetone, separated via centrifugation, and further purified by an ethanol–acetone mixture for three times.

Second growth of M (M = Au, Rh, Cu, Ag, Fe) using concave Pt$_3$Ni as seeds. As a similar procedure, the growth of M (M = Au, Rh, Cu, Ag, Fe) were conducted by simply replacing the Ni(acac)$_2$ by other metal salts. The 4 mg sample of concave Pt$_3$Ni alloy seeds was diluted by 5 mL benzoic alcohol to 0.8 mg/mL (containing 0.0185 mmol of Pt and 0.0064 mmol Ni base on ICP-MS measurement) and sonicated for 5 min. The salts of M were HAuCl$_4$.4H$_2$O, Rh(acac)$_3$, Cu(acac)$_2$, AgNO$_3$, and Fe(acac)$_3$, respectively. Then, the solutions containing M (with the amount from 0.017 mmol to 0.0064 mmol to 0.052 mmol depending on the needs) were added into the 5 mL benzyl alcohol together with 0.05 g benzoic acid and 0.05 g PVP and further stirred for 10 min. The resulting turbid liquid was transferred to a 12-mL Teflon-lined stainless-steel autoclave. The sealed vessel was then heated at 150 °C for a 12 h. When it was cooled down to room temperature, the products were first precipitated by excess acetone, separated via centrifugation, and further purified by an ethanol–acetone mixture for three times.

Typical procedure for the Suzuki-Miyaura cross coupling reaction. In a 10-ml round flask, the catalysts (Pt$_3$Ni or Pt$_3$Ni@Au, both contain 0.005 mmol Pt, 0.5 atom%) and 1.2 mmol of arylboronic acid (purchased from Alfa Aesar) and 3 equiv. (3 mmol) of K$_3$PO$_4$ were added to a solution (3 ml) of H$_2$O/ethanol (v/v = 3:2) with magnetic stirring. Then, 1 mmol 4-bromoanisole (99 %, Alfa Aesar) was dropped into the reaction vial. The temperature of reaction was kept at 30 °C until the reaction stopped. The product was subsequently characterized by GC and 1H NMR.

Typical procedure for hydrogenation of nitrobenzene using the formic acid as hydrogen source. In a 10-ml round flask, 51 μL nitrobenzene (0.5 mmol), 3 equiv. formic acid (1.5 mmol), and the catalysts (Pt$_3$Ni or Pt$_3$Ni@Au, both contain

0.0025 mmol Pt, 0.5 atom%) were dissolved in 2.5 mL ethyl acetate. The round flask was purged with with N_2. Then, the reaction mixture was stirred at 70 °C under a N_2 balloon. The product was subsequently characterized by GC and 1H NMR.

Typical procedure of gas-phase selective oxidation of benzoic alcohol. The gas-phase selective oxidation of alcohols on these catalysts with molecular oxygen was carried out on a fixed-bed quartz tube reactor (inner diameter, 16 mm) under atmospheric pressure. Alcohols were fed continuously using a high-performance liquid pump (weight hourly space velocity (WHSV) was 20 h^{-1}), in parallel with N_2 as the shielding gas, into the reactor heated to the desired reaction temperature. The effluent was cooled using an ice–salt bath to liquefy the condensable vapors for analyzing the productivity and percent conversion by an SP-6890 gas chromatography-flame ionization detector (GC-FID).

4.3 Results and Discussion

4.3.1 Second Growth of Ni and Shape Recovery of Nanocrystals

Through selective etching of the more active metal and rearrangement of the remaining metal atoms, we demonstrated that chemical etching can be utilized as an effective strategy to excavate surfaces of bimetallic NCs and thus generate defects [16]. Figure 4.1 showed the spherical aberration (SA)-corrected high-resolution TEM (HRTEM) image of the concave Pt_3Ni alloys NCs by etching the

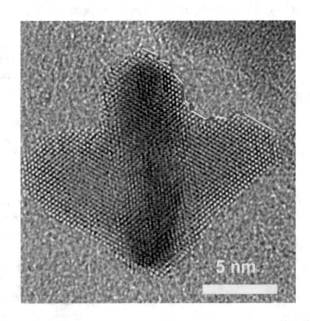

Fig. 4.1 The HRTEM image with spherical aberration correction of an individual concave Pt_3Ni seed contained rich steps and defects. Reprinted with the permission from Ref. [17]. Copyright 2013 American Chemical Society

5 nm

PtNi$_3$ octahedrons, whose surfaces possessed a high density of concave atomic steps (ASs) that was formed by the low-coordinate atoms.

After we obtained such a structure, the concave Pt$_3$Ni seeds were redispersed in a Ni^{2+}-rich chemical solution and underwent a solvothermal process again. From a thermodynamic point of view, the γ{111} facets have the lowest surface energy [18]. As we predicted, the octahedral structures were reborn after the seeded growth process of Ni. Additional HRTEM images showed that the octahedral structures were bounded by eight well-defined {111} facets. To further identify the microstructure of the regained octahedrons, elemental mapping measurements were carried out. For the original concave Pt$_3$Ni seeds, both Pt and Ni were evenly distributed in each nanoparticle, indicating the alloyed phase of concave Pt$_3$Ni (Fig. 4.2).

Quite strikingly, in the as-obtained octahedral NCs, Pt exhibited a hexapod-like distribution as in the initial concave octahedrons while Ni had a uniform distribution in the octahedrons. Thus, there may be a shell of Ni grown outside the core of concave octahedral Pt$_3$Ni, resulting in the formation of the octahedral core–shell structure. The cross-sectional compositional line profiles of Pt$_3$Ni@Ni oriented along different directions also corroborated this unconventional core–shell architecture. For comparison, we also used regular Pt$_3$Ni octahedrons as the seeds for the second growth of Ni and observed an uneventful shape evolution that was confined to the size enlargement. X-ray diffraction (XRD) has proven to be a facile and versatile technique to capture the internal crystal structure changes during the seeded growth process. For only size-enlarged octahedral NCs, the peak positions gradually shifted to higher 2θ values relative to the decreased lattice spacing. Considering that the octahedrons we used as the seeds were entirely bounded by eight most stable {111} faces for facecentered-cubic (fcc) structures, thus, it might follow the diffusion-dominated growth mechanism [19] when we used the octahedral seeds for the second growth of Ni. The smaller Ni atoms would diffuse into the lattice of Pt, which would affect the rearrangement of Pt atoms and the shrink of the lattice. The process of the evolution was concluded in Fig. 4.3. Obviously, the process was completely different when concave Pt$_3$Ni octahedrons were used as the seeds for the second growth of Ni.

The well-defined XRD peaks corresponding to alloyed Pt$_3$Ni remained unchanged, while new emerging peaks could be indexed to fcc nickel grown on the surface of concave Pt$_3$Ni seeds. According to the XRD patterns, we could also unequivocally confirm the Pt$_3$Ni@Ni core–shell structure with an alloyed Pt$_3$Ni core and Ni shell obtained by the second growth of Ni on the concave Pt$_3$Ni octahedrons.

4.3.2 The Surface Defect Effects During the Growth of Nanocrystals

To further elucidate the physical and chemical origin of the shape revolution dominated by the defects, we used density functional theory (DFT) to help us to have a better understanding about the beneficial effects of defects on the shape- and

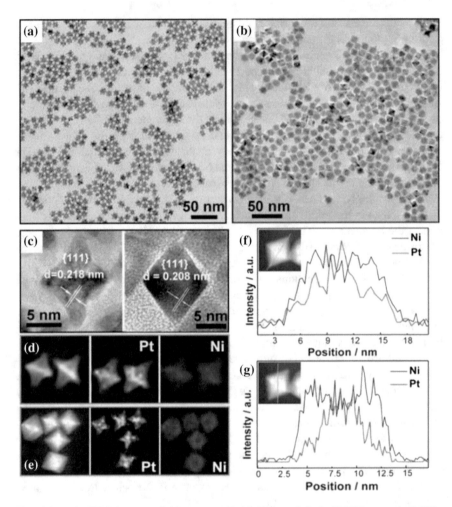

Fig. 4.2 a, b TEM images of (**a**) concave Pt₃Ni NCs and **b** Pt₃Ni@Ni core–shell NCs. **c** HRTEM images of concave (*left*) Pt₃Ni and (*right*) Pt₃Ni@Ni core–shell NCs. **d, e** Elemental maps of (**d**) concave Pt₃Ni and **e** Pt₃Ni@Ni core–shell NCs. **f, g** Cross-sectional compositional line profiles of a Pt₃Ni@Ni core–shell octahedron. Reprinted with the permission from Ref. [17]. Copyright 2013 American Chemical Society

composition-controlled synthesis of multimetallic NCs. Only in this way, we could find the essence from a phenomenon and refine the laws into practical strategies to construct more useful nanostructures. The most prominent characteristic of the concave Pt₃Ni NCs is that they are fully covered with numerous Pt-segregated ASs instead of eight equivalent {111} faces as in the octahedral NCs. According to Fig. 4.4 all of the ASs has two kinds of surface atoms, step-edge (SE) and step-terrace (ST), leading to two different binding sites. In fact, the stable binding of fresh Ni atoms to the ASs initiates the shape recovery of the octahedrons. To this

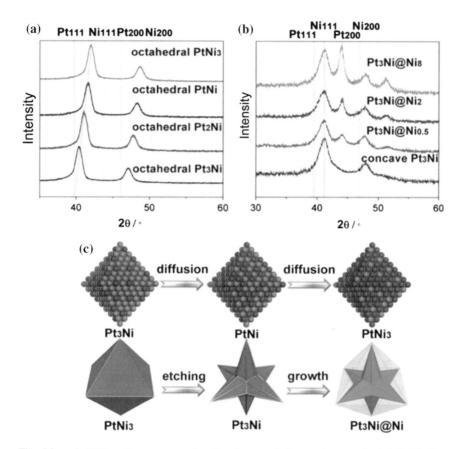

Fig. 4.3 a, b XRD patterns representing the phase evolutions **a** from octahedral Pt₃Ni alloy seeds to octahedral PtNi₃ alloy NCs and **b** from concave PtNi₃ alloy seeds to octahedral Pt₃Ni@Ni core–shell NCs. **c** Schematic illustrations of the two different seeded growth processes. Reprinted with the permission from Ref. [17]. Copyright 2013 American Chemical Society

end, we first compared the stabilities of Ni atoms located at the different sites on the {111} ASs. The DFT calculations indicated that Ni atoms prefer the SE sites, where they are simultaneously coordinated by two SE Pt atoms and three ST Pt atoms. Thus, the deposition of Ni on the defect sites can be viewed as the edge-dependent continuous deposition of Ni on the {111} surface. In this manner, we can affirm that the shape recovery follows a distinctive "step-induced/terrace-assisted" growth mechanism. Further generalizing to other metals akin to Ni, we found that the growth of M (M = Au, Ag, Cu, Rh, Fe) proceeds by "step-induced/terrace-assisted" growth mechanism. More precisely, the step-induced behavior occurs regardless of the crystal structure and lattice parameter of M because of its growth along the [110] direction, aligned with the {111} SE of the Pt₃Ni NCs. However, the subsequent terrace-assisted growth is strongly dependent on the structural and electronic properties of M relative to Pt as the interface forms.

(a)

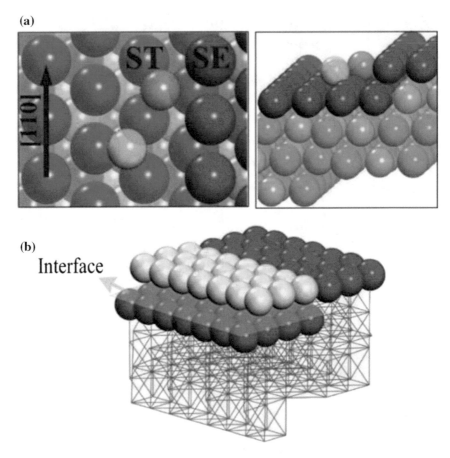

Fig. 4.4 Atomic structures of the step-edge and step-terrace of concave Pt₃Ni nanocrystals: **a** top view and side view, respectively. **b** Interface between M atoms and concave Pt₃Ni nanocrystals during shape recovery. Red, blue, and yellow balls/lattice sites represent Pt, Ni and M atoms, respectively. Reprinted with the permission from Ref. [17]. Copyright 2013 American Chemical Society

Similar to the most Pt-based bimetallic materials, the Pt₃Ni–M interface effects were expected to be small, if M has the same crystal structure (fcc) and only small differences in electron distribution at the interface [20], when the third metal M (M = Au, Ag, Cu, Rh, Ni) deposited on the interface. In contrast, for M=Fe, which has a body-centered-cubic (bcc) structure and a significantly different electron distribution due in large part to its different coordination environment, the lattice mismatch [bcc(110) vs. fcc(111)] at the interface should result in a higher interface energy during the process of the shape revolution, which should hamper the recovery growth [21] (Table 4.1).

Table 4.1 Binding energies of M (M = Au, Ag, Cu, Rh, Ni, Fe) on the {111} Pt_3Ni step (in eV/atom)

Entry	Metal	Eb, ST	Eb, SE	Atomic radius	Crystal structure
1	Au	−2.35	−3.11	1.79	fcc
2	Ag	−2.09	−2.68	1.75	fcc
3	Cu	−2.79	−3.55	1.57	fcc
4	Rh	−3.81	−4.99	1.83	fcc
5	Fe	−3.79	−4.76	1.72	bcc
6	Ni	−3.71	−4.61	1.62	fcc
7	Pt	−	−	1.83	fcc

The atomic radius (in Å) and crystal structure of M are shown together

4.3.3 Design and Synthesis of the Trimetallic Core–Shell Pt–Ni@M Catalysts

To reinforce the "step-induced/terrace-assisted" growth mechanism, we further carried out the consistent synthesis of trimetallic NCs based on concave Pt_3Ni seeds. Figure 4.5 summarizes the growth process of Au deposited on the surface of concave octahedral Pt_3Ni. We could obtain a series of Pt_3Ni@Au trimetallic core–shell structures with different degrees of concavity and surface Au coverages by controlling the amount of chloroauric acid. This mechanism favors achieving shape and composition control of trimetallic nanostructures with perfect monodispersity. Consistent with our theoretical predictions, the Au atoms or clusters would first deposit on the atoms of surface defects. The SA-corrected HRTEM image of Pt_3Ni@$Au_{0.5}$ shows that the atomic steps remained omnipresent after trace deposition of Au and that the concave morphology was still predominant for the products. However, the Pt atomic steps gradually vanished with the increased deposition of Au atoms and the newly generated Au atomic steps served as new preferential nucleation sites. Finally, this type of selective-deposition growth model would confine the Au atoms around the defects when the perfect {111} faces were repaired, generating the octahedral structures and ending the growth.

Elemental mapping was also employed to track this Au growth process. During the gradual growth process, the concave octahedron-like distributions of Pt and Ni continuously remained the same as the Pt_3Ni seeds used in the original stage, maintaining the distinctive environment invariable inside the core–shell structures. For Au, a cruciform area with darker contrast was conspicuously visible in the middle of the NCs, evidencing a saddle-shaped distribution of Au. In the XRD patterns, a set of peaks assigned to fcc Au gradually emerged beside the peaks belonging to Pt_3Ni alloy with the addition of Au (Fig. 4.6). X-ray photoelectron spectroscopy (XPS), which is sensitive to the surface, showed an elevated surface atomic ratio of Au resulting from the growth of the Au on the surface, also testifying to the segregation of Au on the surface (Fig. 4.7). Thus, the combination of HRTEM, elemental mapping, XRD, and XPS confirmed that the Au was

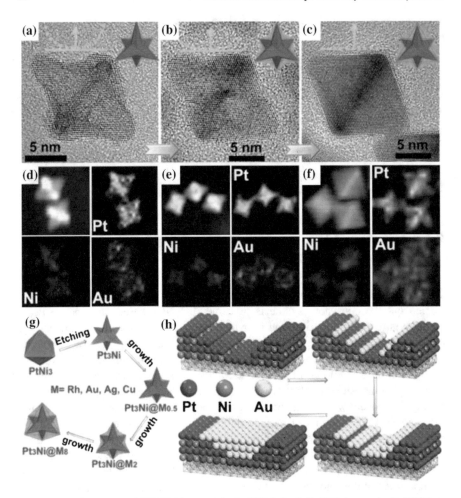

Fig. 4.5 **a–c** SA-corrected HRTEM images of **a** Pt$_3$Ni@Au$_{0.5}$, **b** Pt$_3$Ni@Au$_2$, and **c** Pt$_3$Ni@Au$_8$. **d–f** Elemental maps of **d** Pt$_3$Ni@Au$_{0.5}$, **e** Pt$_3$Ni@Au$_2$, and **f** Pt$_3$Ni@Au$_8$. **g** Schematic illustration of the evolution from octahedral PtNi$_3$ to Pt$_3$Ni@M$_8$ (M = Rh, Au, Ag, Cu). **h** Schematic illustration of the growth of Au on the Pt surface. Reprinted with the permission from Ref. [17]. Copyright 2013 American Chemical Society

preferentially added to the defect sites on the surface during growth, finally leading to these unique trimetallic octahedral core–shell structures (Table 4.2).

To further understand the critical role of defects during the growth process, we subsequently carried out two control experiments. On the one hand, adding more Au precursors after the formation of the octahedrons did not result in bigger octahedrons or particles with uniform shape, but gave mixed morphologies, indicating that the defects played a crucial role in refining the thermodynamically controlled growth pathway of the third metal (Fig. 4.8). On the other hand, putting the Pt, Ni,

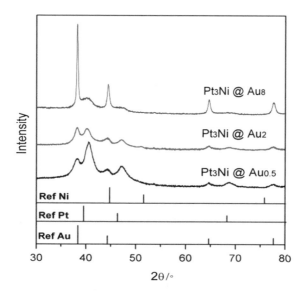

Fig. 4.6 XRD patterns showing the phase evolution process of Au grow on the surface of concave Pt_3Ni and finally form $Pt_3Ni@Au$ core–shell structures. Reprinted with the permission from Ref. [17]. Copyright 2013 American Chemical Society

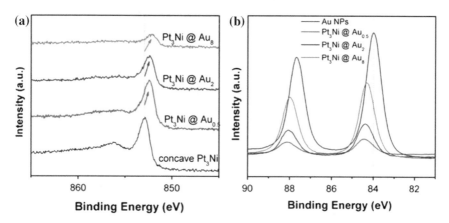

Fig. 4.7 **a** High-resolution scans of the Ni 2p XPS data for the initial concave octahedral Pt_3Ni and $Pt_3Ni@Au$ core–shell NCs. **b** High-resolution scans of the Au 4f XPS data for the as-prepared octahedral PVP-capped Au NCs and octahedral $Pt_3Ni@Au$ core–shell NCs. Reprinted with the permission from Ref. [17]. Copyright 2013 American Chemical Society

and Au precursors into the reactor together, and carrying out a one-pot synthesis under the similar conditions but without the concave Pt_3Ni seeds afforded only alloyed Pt–Ni–Au NPs with irregular shapes (Fig. 4.9). Consistent with the theoretical prediction, the seeded growth of iron with its bcc structure did not lead to the octahedral core–shell $Pt_3Ni@Fe$ structures (Fig. 4.10).

Table 4.2 Surface composition and BE peaks of as-prepared octahedral trimetallic $Pt_3Ni@Au$ nanostructures

Entry	Catalysts	4f7/2 peak of Pt	2p3/2 peak of Ni	4f7/2 peak of Au	Analyzed atomic ratios (Pt:Ni:Au)
1	Concave Pt_3Ni	71.6	852.9		0.76: 0.24:0
2	$Pt_3Ni@Au_{0.5}$	71.3	852.6	84.5	0.62:0.21:0.17
3	$Pt_3Ni@Au_2$	71.2	852.5	84.4	0.37:0.8:0.55
4	$Pt_3Ni@Au_8$	71.1	852.3	84.3	0.25:0.05:0.70
5	Au NCs			84.0	

Fig. 4.8 TEM image of NPs obtained by seeds growth of Au with an excess quantity of chloroauric acid (molar ratio of Pt:Au is 1:6). Reprinted with the permission from Ref. [17]. Copyright 2013 American Chemical Society

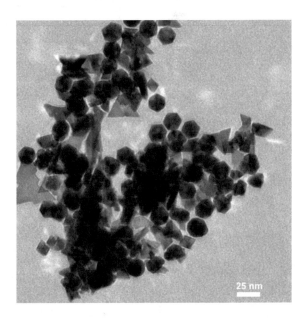

Combining the DFT calculations, this defect-dominated strategy was readily generalized to the growth of other metals, such as Ag, Cu, and Ru, on the surface of the concave octahedral Pt_3Ni, generating the distinctive trimetallic octahedral core–shell structures. All of these trimetallic structures could be further verified by XRD and energy-dispersive spectroscopy (EDS) measurements (Fig. 4.11).

4.3.4 The Study of the Catalytic Properties of the Trimetallic Core–Shell Pt–Ni@M Catalysts

By sophisticated decoration with the third metal on the surface of bimetallic NCs, the geometrical and electronic effects of the NCs could be fine tuned to optimize the catalytic activity and selectivity.

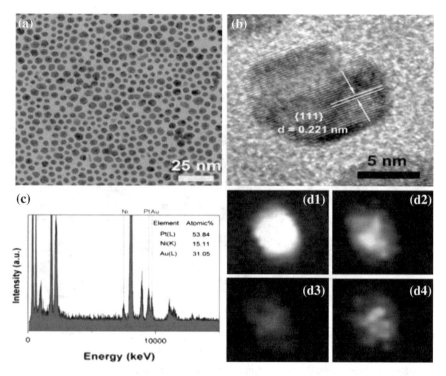

Fig. 4.9 **a** TEM image, **b** HRTEM image, and **c** EDX patterns of Pt–Ni–Au NPs synthesized by one-pot co-reduction (molar ratio of precursors: Pt:Ni:Au = 3:1:3) **d1** HAADF-STEM image, and corresponding elemental maps of **d2** Pt (L), **d3** Ni (K), **d4** Au (L). Reprinted with the permission from Ref. [17]. Copyright 2013 American Chemical Society

To the best of our knowledge, Au(0) is inert with respect to activation of C–X bonds (X = Cl, Br, I) [22]. According to the rendered mechanism, we could hypothesize that Pt(0) is the active state for the Suzuki-Miyaura reaction. The traditional concept believed that the valence state of Pt would increase in the oxidative addition process and decrease to Pt(0) in the subsequent reductive elimination during the Suzuki-Miyaura reaction, generating a complete catalytic cycle. From Fig. 4.12, the catalytic activity of $Pt_3Ni@Au_{0.5}$ catalyst was exceptionally higher than Pt_3Ni catalyst, probably because of the presence of trace Au on the surface, which might facilitate the reduction of Pt(II) to Pt(0) and favor the formation of the catalytic active species. Surprisingly, the activity of the trimetallic catalyst with increasing quantities of Au exhibited a significant decay rather than an increasing trend, which may result because the excess deposited Au blocks the exposed Pt active sites, breaking the interaction with substrate (Figs. 4.13 and 4.14).

Some recent progress in catalysis of dehydrogenation reactions of formic acid (FA) indicated that FA can be used as a safe and convenient hydrogen carrier for energy applications [23], because FA is one of the major products in biomass processing. H_2 is released via the catalytic dehydrogenation reaction

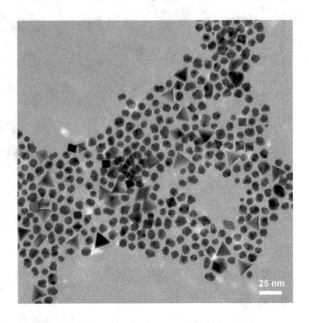

HCOOH → H$_2$ + CO$_2$, but is usually accompanied by low levels of CO produced
by the side reaction HCOOH → H$_2$O + CO. Utilizing the stoichiometric hydrogenation of nitrobenzene to aniline, we could establish a tandem reaction to
evaluate the activity and selectivity of the decomposition of FA. Among the as-
prepared trimetallic multifunctional catalysts, Pt$_3$Ni@Au$_{0.5}$ (containing 0.5 atom%
Pt) exhibited the best performance with respect to both activity and selectivity, and a similar deactivation trend as the Suzuki-Miyaura reaction was also
observed when too much Au was grown on the surface of the concave Pt$_3$Ni seeds
(Table 4.3). Thus, the presence of the atomic steps in the trimetallic catalysts plays
a crucial role in maintaining a higher catalytic activity. Notably, the bimetallic surface should be decorated with different metals for specific reactions, which can
achieve the optimization with respect to both catalytic activity and selectivity. As
another comparable important industrial example, Pt$_3$Ni@Ag$_{0.5}$ achieved the best
catalytic activity and selectivity in the selective oxidation of benzyl alcohol to benzaldehyde [24] (Table 4.4).

In this context, Pt$_3$Ni@Au$_{0.5}$ exhibited the best performance for the selective
oxidation of benzyl alcohol to benzaldehyde. To further understand the structure–
activity relationships of trimetallic NCs, this chapter studied the electronic properties of Pt$_3$Ni@Au as a function of the Au coverage from both the experimental
and theoretical perspective. Recently, the observation that electrons can be transferred from the Au domains to Pt in a hybrid structure has been reported [25].
In this work, XPS was also utilized to analyze the electron transfer process and
direction. According to Fig. 4.12c, the deposition of Au on the bimetallic surface
led to an apparent shift of the Pt 4f7/2 and 4f5/2 peaks to lower binding energy
for Pt$_3$Ni@Au relative to concave Pt$_3$Ni, which indicated that the Au atoms may

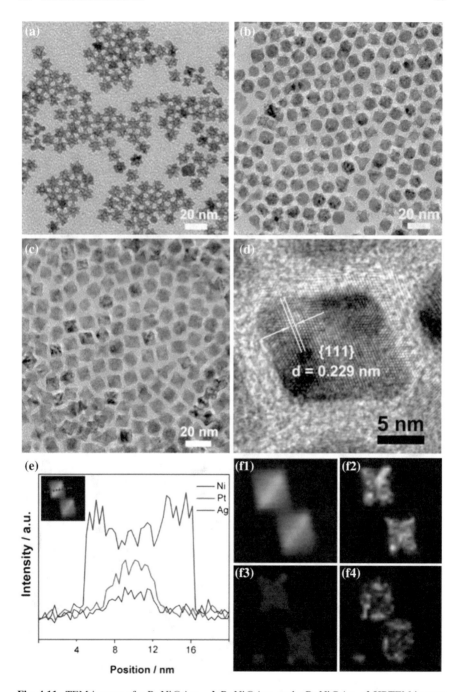

Fig. 4.11 TEM images of **a** $Pt_3Ni@Ag_{0.5}$, **b** $Pt_3Ni@Ag_2$, and **c** $Pt_3Ni@Ag_8$. **d** HRTEM images of $Pt_3Ni@Ag_8$. **e** EDS line profiles of an individual $Pt_3Ni@Ag_8$ nanoparticle. **f1** HAADF-STEM image, and corresponding elemental maps of **f2** Pt (L), **f3** Ni (K), **f4** Ag (L). Reprinted with the permission from Ref. [17]. Copyright 2013 American Chemical Society

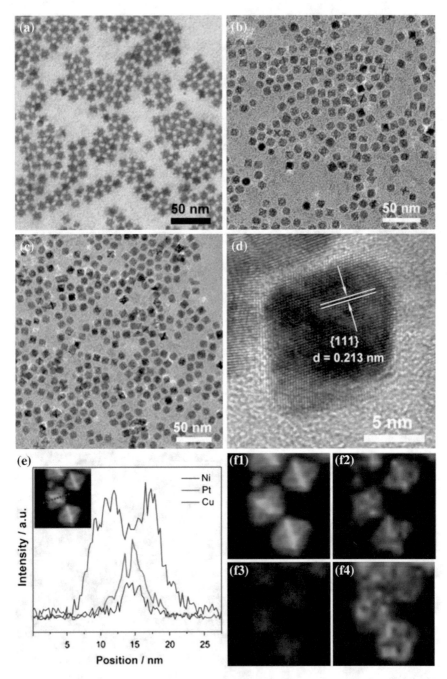

Fig. 4.12 TEM images of **a** Pt$_3$Ni@Cu$_{0.5}$, **b** Pt$_3$Ni@Cu$_2$, and **c** Pt$_3$Ni@Cu$_8$. **d** HRTEM images of Pt$_3$Ni@Cu$_8$. **e** EDS line profiles of an individual Pt$_3$Ni@Cu$_8$ nanoparticle. **f$_1$** HAADF-STEM image, and corresponding elemental maps of **f$_2$** Pt (L), **f$_3$** Ni (K), **f$_4$** Cu (K). Reprinted with the permission from Ref. [17]. Copyright 2013 American Chemical Society

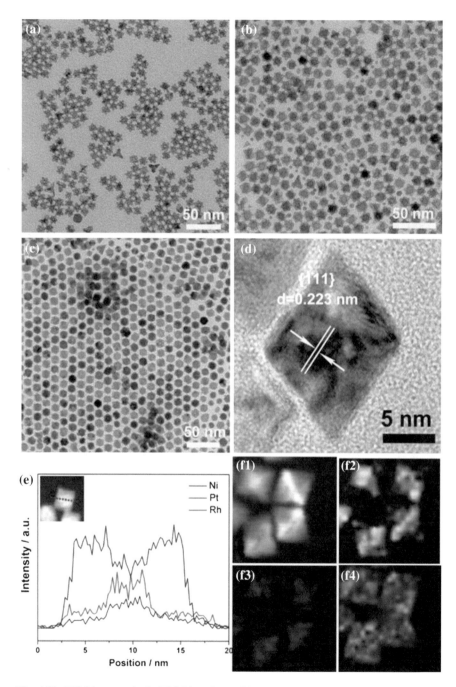

Fig. 4.13 TEM images of **a** $Pt_3Ni@Rh_{0.5}$, **b** $Pt_3Ni@Rh_2$, and **c** $Pt_3Ni@Rh_8$. **d** HRTEM images of $Pt_3Ni@Rh_8$. **e** EDS line profiles of an individual $Pt_3Ni@Rh_8$ nanoparticle. **f₁** HAADF-STEM image, and corresponding elemental maps of **f₂** Pt (L), **f₃** Ni (K), **f₄** Rh (L). Reprinted with the permission from Ref. [17]. Copyright 2013 American Chemical Society

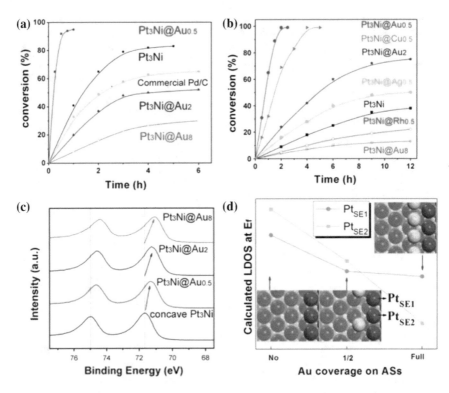

Fig. 4.14 Conversion as a function of time in **a** the Suzuki–Miyaura reaction and **b** reduction of nitrobenzene using formic acid as a hydrogen source catalyzed by trimetallic catalysts containing 0.5 atom%Pt. **c** High-resolution Pt 4f XPS scans for the initial concave octahedral PtNi$_3$ and tri-metallic Pt$_3$Ni@Au NCs. **d** DFT-calculated local densities of states (LDOS) at the Fermi energy (E$_F$) for Pt$_{SE1}$ and Pt$_{SE2}$ atoms as functions of the Au coverage at the atomic steps. The atomic structures are shown in the insets. *Red*, *blue*, and *yellow* balls respresent Pt, Ni, and Au atoms, respectively. Reprinted with the permission from Ref. [17]. Copyright 2013 American Chemical Society

donate electrons to Pt, thereby reducing the oxidized Pt(II). Similarly, the electron transfer process also existed between the Au and Ni. On the other hand, the DFT calculations clearly indicated that the deposition of Au on the Pt step apparently reduces the local density of states (LDOS) at the Fermi energy (EF) for Pt atoms at the ASs due to the d10 electron configuration of Au. According to the Anderson-Newns model for adsorbate-metal bonding [26], such a decrease means that Au-deposited Pt ASs have fewer electronic states to respond to the adsorbate, decreasing their ability to facilitate the interaction between the adsorbate and the NC. It is expected that these two opposite process would make mixed Au–Pt ASs to achieve the optimization of the catalytic properties. From the experimental perspective, we indeed found a great structure with optimized ratio of Au and Pt, achieving the highest activity and selectivity. Furthermore, the deposition of Au

Table 4.3 Turnover frequencies (TOFs [$h-1$]) of the hydrogenation of nitrobenzene catalyzed by various catalysts

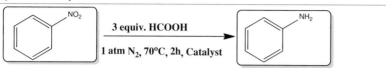

Entry	Catalyst	Conversion (%)	Yield (%)	TOFs ($h-1$)
1	Octahedral PtNi$_3$	5	5	9
2	Concave Pt$_3$Ni	9	8	12
3	Pt$_3$Ni@Au$_{0.5}$	>99	97	124
4	Pt$_3$Ni@Au$_2$	24	23	86
5	Pt$_3$Ni@Au$_8$	4	4	6
6	Pt$_3$Ni@Ag$_{0.5}$	16	15	18
7	Pt$_3$Ni@Cu$_{0.5}$	69	67	74
8	Pt$_3$Ni@Rh$_{0.5}$	5	5	8

Table 4.4 Turnover frequencies (TOFs [$h-1$]) of the hydrogenation of nitroarenes catalyzed by various catalysts

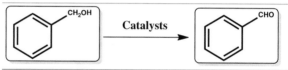

Catalyst	Temp. (°C)	Conversion (%)	Selectivity (%)
Pt$_3$Ni@Ag$_{0.5}$	210	92.1	96.2
	230	96.2	96.1
	250	98.4	96.0
Pt$_3$Ni@Ag$_2$	210	85.9	91.2
	230	93.4	91.4
	250	97.1	91.0
Pt$_3$Ni@Ag$_8$	210	63.3	96.5
	230	74.1	96.5
	250	92.4	96.5
Pt$_3$Ni@Rh$_{0.5}$	230	43.7	84.8
	250	30.9	62.0
	280	30.2	73.9
Pt$_3$Ni@Au$_{0.5}$	230	12.3	95.1
	250	12.1	95.6
	280	13.4	96.1
Pt$_3$Ni@Cu$_{0.5}$	230	90.7	95.0
	250	91.6	91.3
	280	85.9	90.2

on the Pt would to a certain extent control the type of the reactive sites and further affect the selectivity of the catalysis. In a word, the introduction of the third metal may open the way to design new catalysts with controllable reaction activity and selectivity. In particular, the step-induced growth mechanism proposed in our work allows the controllable synthesis of the surface formed by various metals.

4.4 Conclusions

In this chapter, we have discovered a novel shape recovery phenomenon of NCs that is dominated by the defects existed in the growth process of the NCs. On the basis of the rendered "step-induced/terrace-assisted" growth mechanism, we une-quivocally confirmed two requirements for the synthesis of a new trimetallic or multimetallic core–shell structure by shape recovery: (1) the same crystal structure of the third metal relative to the original two metals; (2) similar atomic radius of the third metal relative to the original two metals. Remarkably, the catalytic per-formance can be significantly improved by sophisticated decoration with the third metal on the bimetallic surface. We foresee that this method will greatly expand our abilities to design a new nanocrystal catalysts with fine-tuned geometric and electronic structures.

References

1. Hong JW, Kang SW, Choi B-S, Kim D, Lee SB, Han SW (2012) Controlled synthesis of Pd–Pt alloy hollow nanostructures with enhanced catalytic activities for oxygen reduction. ACS Nano 6(3):2410–2419
2. Guo S, Wang E (2011) Noble metal nanomaterials: controllable synthesis and application in fuel cells and analytical sensors. Nano Today 6(3):240–264
3. Gao C, Lu Z, Liu Y, Zhang Q, Chi M, Cheng Q, Yin Y (2012) Highly stable silver nanoplates for surface plasmon resonance biosensing. Angew Chem Int Ed 51(23):5629–5633
4. Zhou K, Li Y (2012) Catalysis based on nanocrystals with well-defined facets. Angew Chem Int Ed Engl 51(3):602–613
5. Quan Z, Wang Y, Fang J (2012) High-Index faceted noble metal nanocrystals. Accounts Chem Res 46(2):191–202
6. Zhang L, Niu W, Xu G (2012) Synthesis and applications of noble metal nanocrystals with high-energy facets. Nano Today 7(6):586–605
7. Dahl S, Logadottir A, Egeberg R, Nielsen JH, Chorkendorff I, Törnqvist E, Nørskov JK (1999) Role of steps in N-2 activation on Ru (0001). Phys Rev Lett 83(9):1814–1817
8. Liu Z-P, Hu P (2003) General rules for predicting where a catalytic reaction should occur on metal surfaces: a density functional theory study of CH and CO bond breaking/making on flat, stepped, and kinked metal surfaces. J Am Chem Soc 125(7):1958–1967
9. Stamenkovic V, Mun BS, Mayrhofer KJJ, Ross PN, Markovic NM, Rossmeisl J, Greeley J, Norskov JK (2006) Changing the activity of electrocatalysts for oxygen reduction by tuning the surface electronic structure. Angew Chem Int Ed 45(18):2897–2901
10. Liu J, Chenna S, Crozier PA, Li Y, Chen A, Shen W (2012) Stabilized gold nanoparticles on Ceria nanorods by strong interfacial anchoring. J Am Chem Soc 134(51):20585–20588

11. Rodriguez J, Ma S, Liu P, Hrbek J, Evans J, Perez M (2007) Activity of CeOx and TiOx nanoparticles grown on Au (111) in the water-gas shift reaction. Science 318(5857):1757–1760

12. Wang DY, Chou HL, Lin YC, Lai FJ, Chen CH, Lee JF, Hwang BJ, Chen CC (2012) Simple replacement reaction for the preparation of ternary Fe1−x PtRu x nanocrystals with superior catalytic activity in methanol oxidation reaction. J Am Chem Soc 134(24):10011–10020

13. Wanjala BN, Fang B, Luo J, Chen Y, Yin J, Engelhard MH, Loukrakpam R, Zhong CJ (2011) Correlation between atomic coordination structure and enhanced electrocatalytic activity for trimetallic alloy catalysts. J Am Chem Soc 133(32):12714–12727

14. Xia Y, Xiong Y, Lim B, Skrabalak SE (2009) Shape-controlled synthesis of metal nanocrystals: simple chemistry meets complex physics? Angew Chem Int Ed 48(1):60–103

15. Zhou ZY, Tian N, Li JT, Broadwell I, Sun SG (2011) Nanomaterials of high surface energy with exceptional properties in catalysis and energy storage. Chem Soc Rev 40(7):4167–4185

16. Wu Y, Wang D, Niu Z, Chen P, Zhou G, Li Y (2012) A strategy for designing a concave Pt–Ni alloy through controllable chemical etching. Angew Chem 124(50):12692–12696

17. Wu Y, Wang D, Chen X, Zhou G, Yu R, Li Y (2013) Defect-dominated shape recovery of nanocrystals: a new strategy for trimetallic catalysts. J Am Chem Soc 135(33):12220–12223

18. Wang Z (2000) Transmission electron microscopy of shape-controlled nanocrystals and their assemblies. J Phys Chem B 104(6):1153–1175

19. Chen W, Yu R, Li L, Wang A, Peng Q, Li Y (2010) A seed-based diffusion route to monodisperse intermetallic CuAu nanocrystals. Angew Chem Int Ed Engl 49(16):2917–2921

20. De Boer FR, Boom R, Mattens W, Miedema A, Niessen A (1988) Cohesion in metals: transition metal alloys. Elsevier Science Publishers B V, vol 1 p 758

21. Somoza J, Gallego L, Rey C, Fernandez H, Alonso J (1992) Glass formation in the Cu–Ti–Zr system and its associated binary systems. Philos Mag B 65(5):989–1000

22. Wegner HA, Auzias M (2011) Gold for C-C coupling reactions: a Swiss-Army-knife catalyst? Angew Chem Int Ed Engl 50(36):8236–8247

23. Zhang S, Metin O, Su D, Sun S (2013) Monodisperse AgPd alloy nanoparticles and their superior catalysis for the dehydrogenation of formic acid. Angew Chem Int Ed Engl 52(13):3681–3684

24. Mallat T, Baiker A (2004) Oxidation of alcohols with molecular oxygen on solid catalysts. Chem Rev 104(6):3037–3058

25. Yang J, Ying JY (2011) Nanocomposites of Ag2S and noble metals. Angew Chem Int Ed 50(20):4637–4643

26. Newns D (1969) Self-consistent model of hydrogen chemisorption. Phys Rev 178(3):1123

Chapter 5
Sophisticated Construction of Au Islands on Pt–Ni: An Ideal Trimetallic Nanoframe Catalyst

5.1 Introduction

A lower coordination number, such as corner and edge atoms, are generally the active sites to activate most important catalytic reactions. If the catalysts we designed do not contain the extra atom and all of the useful atoms remain in active sites, then this catalyst is an ideal model catalyst. Here we have developed a priority-related chemical etching method to transfer the starting Pt–Ni polyhedron to a nanoframe. Utilizing the lower electronegativity of Ni in comparison to Au atoms, further in conjunction with the galvanic replacement of catalytic active Au to Ni tops, and then a unique Au island on Pt–Ni trimetallic nanoframe is achieved. The open trimetallic nanoframe holds 36 Pt-segregated edges and 24 Au corners, in which all diverse active sites are connected, allowing the strong synergy effect between Au-corner and Pt–Ni edge. As an ideal model catalyst for the study of structure–activity relationship, the trimetallic nanoframe owning versatile reactive sites shows better activity and the selectivity (100 %) towards the selective hydrogenation of 4-nitrobenzaldehyde, as compared with bimetallic Pt–Ni nanoframe catalysts. Moreover, this Pt–Ni catalysts has outstanding durability (beyond 3000 cycles) of electro-oxidation of methanol. Moreover, the design strategy is based on structural priority mechanism of multimetallic nanocrystals during the synthesis, and thus can be generalized to other analogous metal-bimetallic nanocrystal combinations, such as Pd, Cu islands on Pt–Ni nanoframes. Thereby, this method is expected to pave the way for future development of efficient catalysts.

As to a polyhedron having a lower coordination number, corner and edge atoms, rather than those at the flat surface, are generally the preferred sites to activate most important catalytic reactions. Because these corner and edge atoms can effectively reduce the activation energy of the reaction and are capable of regulating the adsorption of the reaction substrate on the catalyst surface [1, 2]. Nanostructures with open surface features such as high-index facet [3], concave surface [4], and

© Springer-Verlag Berlin Heidelberg 2016
Y. Wu, *Controlled Synthesis of Pt–Ni Bimetallic Catalysts and Study of Their Catalytic Properties*, Springer Theses,
DOI 10.1007/978-3-662-49847-7_5

porous structure have provoked ever-increasing attention in the field of catalysis according to their exceptional catalytic properties. Metallic NPs with a nanoframe structure, in which edges are joined together at specific corners to create a skeletal frame, can harvest the costly materials and supply the three-dimensional molecular accessibility, thus serving as an ideal model catalyst to satisfy the ultimate goal of nanoparticle (NP) catalysis. Several pioneering works have been established to obtain this unique nanostructure with outstanding physical and chemistry properties, such as monometallic Au [5, 6], bimetallic Au–Ag [7] and Pt–Cu nanoframes [8]. Chen et al. observed that a Pt_3Ni rhombic dodecahedral nanoframe composed of 24 edges and 14 corners enable the maximal utilization of Pt by a three-dimensional accessible surface [9]. The fascinating geometrical and electronic effects induced by the nano-segregated Pt-skin surface confer a substantial improvement in activity towards the oxygen reduction reaction to this unique nanoframe catalyst. Encouraged by these progresses, designing the multicenter and multifunctional nanostructures with high-efficiency is currently becoming one promising area.

In this chapter, a previously unknown sequential strategy is exploited to construct novel trimetallic nanoframe catalyst. The corners of this nanoframe can be sophisticatedly anchored with Au islands by galvanic replacement, masterly utilizing the highly active top Ni atoms, and this displacement reaction takes advantage of the difference in the activity of high active Ni atoms on the corners and Au atoms of high electronegativity. In the sequential process, structural priority-related chemical etching is employed to carefully regulate the corrosion of Ni and rearrangement of Pt atoms by utilizing the different reactivity of Ni atoms in different geometric sites. This ingenious design not only guarantees the major proportion of versatile active sites, but also directly supports the electron transfer process of the heterogeneous structure in different species, which will extend our ability to improve activity, selectivity, and durability. We believe that such previously unreported synthetic strategies, integrating selective galvanic replacement, and structural priority-related chemical etching provides more options for selecting ideal NP catalysts.

5.2 Experimental Section

5.2.1 Materials and Intruments

Chemicals: Analytical grade benzyl alcohol, aniline, chloroauric acid, and dimethylglyoxime were obtained from Beijing Chemical Reagents, China. $Pt(acac)_2$, (99 %) $Ni(acac)_2$ (99 %), and PVP (MW = 8000, AR) were purchased from Alfa Aesar. All of the chemicals used in this experiment were analytical grade and used without further purification.

Characterization: The crystalline structure and phase purity were determined by Rigaku RU-200b X-ray powder diffractometer with CuKa radiation (1 = 1.5418 Å). The composition of the product was measured by the inductively coupled plasma-mass spectrometry (ICP-MS) and energy dispersive spectrometer (EDS). TEM images were recorded by a JEOL JEM-1200EX working at 100 kV. HRTEM images were recorded by a FEI Tecnai G2 F20 S-Twin high-resolution transmission electron microscope working at 200 kV and a FEI Titan 80–300 transmission electron microscope equipped with a spherical aberration (Cs) corrector for the objective lens working at 300 kV. The catalytic reaction result was measured by gas chromatography (GC) (SP-6890) and ^{1}H NMR. The NMR spectroscopy was conducted on a JEOL JNM-ECX 400 MHz instrument.

5.2.2 Experimental Methods

Preparation of truncated octahedral PtNi₃ nanocrystals: In a typical synthesis of PtNi₃ octahedral nanocrystals, Pt(acac)₂, (40.0 mg), poly(vinylpyrrolidone) (PVP, MW = 8000, (400.0 mg) Ni(acac)₂ (75 mg), and 0.5 mL aniline were dissolved in 25 mL of benzylalcohol, followed by 10 min vigorous stirring. The resulting homogeneous green solution was transferred into a 45-mL Teflon-lined stainless steel autoclave. The sealed vessel was then heated at 150 °C for a 12 h before it was cooled down to room temperature. The products were precipitated by acetone, separated via centrifugation and further purified by an ethanol-acetone mixture for three times.

Controlled corrosion of truncated octahedral PtNi₃ nanocrystals: The as-prepared PtNi₃ (4 mg) precursor was redispersed in 1 mL H₂O, followed by the addition of 10 mg dimethylglyoxime dissolved in 1 mL ethanol. Then, the reaction mixture was stirred 12 h. Next, 5 mL acetic acid (50 %) was added and the solution was kept stirred for a further 15 min at room temperature. Finally, the products were obtained via centrifugation and further washed by ethanol.

Preparation of Au-modified Pt–Ni nanoframe: The 2.4 mg sample of concave Pt₃Ni alloy seeds was redispersed in 5 mL H₂O, followed by the addition of 28 μL, 56 μL, or 140 μL the solution of HAuCl₄ and stirred for 0.5 h, followed by the addition of 10 mg dimethylglyoxime dissolved in 1 mL ethanol. Then, the reaction mixture was stirred 12 h. Next, 5 mL acetic acid (50 %) was added and the solution was kept stirred for a further 15 min at room temperature. Finally, the products were obtained via centrifugation and further washed.

Typical procedure for the hydrogenation of nitrobenzene catalyzed by Pt–Ni alloys: First, 0.5 mmol nitrobenzene in ethyl acetate (2.5 mL) and the Pt–Ni alloy or Au on Pt₃Ni catalyst (containing 0.0025 mmol Pt, 0.5 mol%) were added in a 10 mL round bottom flask. The round flask was purged with H₂. Then, the reaction mixture was stirred at room temperature under a H₂ balloon. Then the reaction mixture was maintained at 25 °C in an oil bath. Finally, the product was purified by column chromatography and subsequently characterized by ^{1}H NMR.

Gas-phase selective oxidation of benzyl alcohol: Gas-phase selective oxidation of benzyl alcohol occured in the reactor of a fixed bed quartz tube (inner diameter 16 mm) under the atmospheric pressure. All of the metal catalyst was supported on activated carbon (containing 2 % Pt). In 10 mL round bottom flask, 51 μL (0.5 mmol) nitrobenzene, 3 equiv. formic acid (1.5 mmol), and Pt_3Ni or Pt_3Ni @ Au catalyst (containing 0.0025 mmol Pt, 0.5 mol%) were mixed in 2.5 mL ethyl acetate. Discharging the air in the flask and plugging nitrogen balloon, the mixed solution was reacting in an oil bath kept at 70 °C. The reaction results were characterized by chromatography and NMR.

Electrochemical measurement: 20 mg conductive carbon black was dispersed in 20 mL ethanol in a 100 mL beaker, then 2 mg nanoparticles was added and the solution underwent ultrasonic vibration for 30 min to make the nanoparticles sufficiently dispersed in the conductive carbon black, then the solution was stirred overnight at room temperature. The product was separated via centrifugation and decantation of the supernatant. The black solid obtained was washed by a large amount of deionized water (30 mL × 3) for three times, then was dried, weighed and formulated into an aqueous solution (2 mg/mL).

Polishing electrode: A small amount of Al_2O_3 powder was put on the electrode polishing cloth, with a little distilled water added to make it wet. Then the cleaned electrode was polished by gently and uniformly drawing the number 8 on the cloth, then the electrode was rinsed with distilled water, ultrasonically cleaned by distilled water for 2 s and ethanol for another 2 s, and then rinsed with deionized water. Then the electrode was taken out from the water and shaken to remove the distilled water on its surface. Cyclic voltammogram was scanned in 0.001 M KFe (CN) $_6$–0.1 M KNO_3 mixture at 50 mv/s within the range of +0.60 to −0.20 V, and the difference of reversible peak potential was observed. If the difference is above 80 mV, the electrode surface should be rinsed with distilled water and polished by repeating the previous steps until the difference of reversible peak potential is about 80 mV. After polishing is completed, the electrode was rinsed with deionized water and the water around the glassy carbon electrode was gently wiped off with filter paper.

Pointing sample: 5 μL test solution was dropped on the disc of the glassy carbon working electrode and then evaporated in a fume hood at room temperature. When the solution is about to dry, adding 5 μl of 0.05 % w nafion (purchased 5 % nafion diluted 100-fold) to make the electrode dry naturally.

50 ml electrolyte solution (0.1 M $HClO_4$ or 0.5 M H_2SO_4) was placed in a sealable container (centrifuge tubes or flasks with stoppers). The container was fixed and then purged with N_2 for about 30 min. The hose with a 5 ml pipette tip was connected to a nitrogen bottle, and plugged into the solution and was blown for at least 30 min to make the solution bubble constantly but not splash.

The glassy carbon electrode was used as the working electrode, the Ag/AgCl electrode as the reference electrode and the platinum wire electrode as the counter electrode. Firstly, the acid system was activated by scanning for 50 cycles with a speed of 100 mV/s in the perchloric acid (0.1 M) electrolyte. Then the activity of cyclic voltammetry is measured in the electrolyte solution (potential range: -0.2 V–1.0 V, vs. RHE; scanning speed: 0.05 V/s).

0.6 mL methanol was added in 15 ml as-prepared 0.1 M $HClO_4$ or 0.5 M H_2SO_4 and stirred. Then the solution was scanned for 15 cycles at 50 mV/s. If the peak current is still rising, scan sequentially for another 20–30 cycles until it does not rise any more.

5.3 Results and Discussion

5.3.1 The Corrosion of Truncated Octahedral PtNi₃ Nanocrystals

Using truncated octahedron $PtNi_3$ as the precursor, we employed an effective chemical etching method to prepare Pt–Ni framework successfully [10]. Figure 5.1 shows both transmission electron microscope (TEM) and high-resolution TEM (HRTEM) (inset) images of the truncated octahedral $PtNi_3$ NPs. After dimethylglyoxime was applied to a suspension of the $PtNi_3$ NPs, the Ni could be gradually etched from the $PtNi_3$ alloy by oxidative etching. Figure 5.1b reveals that most of the obtained NPs retained their original symmetry and showed a narrow size distribution (average size of 12.5 ± 1.5 nm). It is proved that NPs obtained by this method has an excellent monodispersity. Examination of the HRTEM image indicates that each particle possesses an eroded hollow frame. The high-angle annular dark-field scanning transmission electron microscope (HAADF-STEM) micrograph indicates that parts of the framework are much brighter than the other region of the particles, further demonstrating the presence of a nanoframe structure. The structural evolution of the particles from the parent $PtNi_3$ polyhedrons to final products was accompanied by a variation in elemental distribution. According to the elemental maps, the distribution of Ni in the starting $PtNi_3$ particles showed a slight enrichment in the central region, whereas a homogeneous distribution of Pt throughout the entire particle was observed.

After chemical etching, the as-obtained truncated octahedral nanoframes retained overall 36 edges and 24 corners of the truncated octahedron, in which both Pt and Ni exhibited homogeneous distributions, implying that the final product formed an alloy phase.

Chemical etching, often involving further carving of the polyhedron precursor, can further control the components and the hole size of the final obtained metal frame. As shown in Fig. 5.2, a gradual increase in the ratio of Ni to Pt (from 1:1 to 2:1, 3:1, 6:1, and 10:1) resulted in a significant increase in the hollowness of the final obtained nanoframes. When the internal cavity became larger, the octahedral symmetry of this framework was also increasingly difficult to maintain. The signal of Pt can be detected through the ICP-MS, which demonstrated Pt elements cannot be etched from the bimetallic system through this method. As the corrosion reaction progressed, the Ni/Pt ratio of nanoparticles decreased unceasingly, which

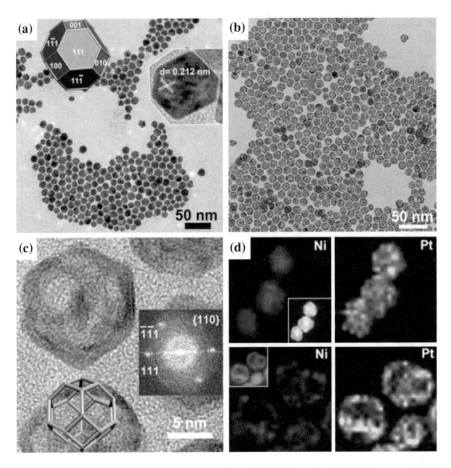

Fig. 5.1 a TEM image of truncated octahedral PtNi₃. **b** TEM and **c** HRTEM image of Pt₃Ni nanoframe (Insets are the fast fourier transform patterns and corresponding ideal model of the nanoframe). **d** HAADF-STEM image and corresponding elemental maps. The top and bottom parts belong to the truncated octahedral PtNi₃ and Pt₃Ni nanoframe, respectively. Reprinted with the permission from Ref. [11]. Copyright 2014 American Chemical Society

made the chemical etching process more difficult. The terminated stable phase of Pt₃Ni can be obtained, which was further confirmed by powder X-ray diffraction (XRD), energy dispersive spectroscopy (EDS), and ICP-MS (Fig. 5.3).

5.3.2 The Analysis of the Etching Mechanism of Truncated Octahedral Pt–Ni NPs

Although the composition of octahedral PtNi₃ and truncated octahedral PtNi₃ is the same, their morphology is different, which leads to different morphology evolution processes through the same etching process. It is undoubtedly very

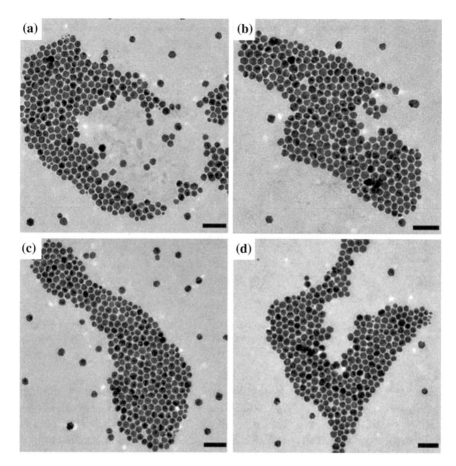

Fig. 5.2 TEM images of **a** truncated octahedral PtNi, **b** truncated octahedral PtNi$_2$, **c** truncated octahedral PtNi$_6$, and **d** truncated octahedral PtNi$_{10}$. Reprinted with the permission from Ref. [11]. Copyright 2014 American Chemical Society

important to understand the etching-dominated fabrication methods of alloy NPs with well-designed shapes. Figure 5.4 summarizes the structure evolution process of octahedral PtNi$_3$ and truncated octahedral PtNi$_3$ after chemical etching. We have thoroughly investigated the changes of the crystal structure, composition and morphology in the etching processes by combining the density functional theory calculations (DFT) with experiments, and attempted to understand the intrinsic mechanisms. A thermodynamic viewpoint showed that the dissolution rate of Ni atoms is faster than that of Pt atoms, following the order edge >(110) ≈ (100) > (111) surfaces. The proposed processes involved in the shape evolution of truncated octahedral PtNi$_3$ NPs are schematically illustrated in Fig. 5.4. As for octahedral PtNi$_3$ NPs [10], the etching of (111) surfaces proceeded in a layer-by-layer removal manner, in which principle yielded a Pt-rich concave

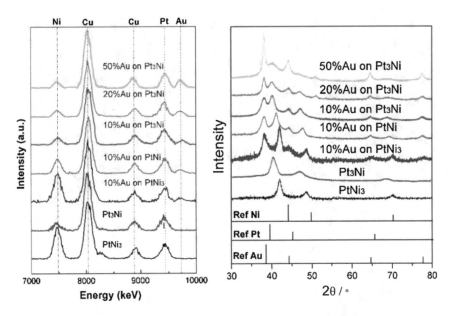

Fig. 5.3 The *left*: EDS patterns of Au on Pt–Ni nanoparticles; the *right*: XRD patterns of Au on Pt–Ni nanoparticles. Reprinted with the permission from Ref. [11]. Copyright 2014 American Chemical Society

solid structure. However, the open (100) surfaces exhibited a distinctly different etching process and mechanism compared to (111) surfaces because of alternate stacking of full Ni and Pt–Ni alternately in the structure. More precisely, after Ni dissolution, the surface Pt aggregated driven by the near-neighbor bonding to form a "porous Pt shell". It inhibited the binding between exposed subsurface Ni and dimethylglyoxime, accordingly, which is detrimental to the occurrence of the Ni atom oxidation corrosion process. The subsequent etching process was unilaterally controlled by outward diffusion of Ni^{2+} from NPs, accompanying the inward flow of vacancies, through the pores of the inert Pt shell. It followed a mechanism analogous to the Kirkendall effect [12, 13]. As the reaction proceeded deeper into the Ni center, the etching rate increased significantly, which allowed mobility of vacancies and growth and coalescence of small cavities. When reaching the Kirkendall hole, the etching of {111} surfaces would open the NPs. Meanwhile, growing stress corrosion cracks would in time decompose the Pt-segregated layers, which was not present in the hollowing of metals [13]. Following the step-induced/terrace-assisted growth mechanism of alloy NPs [14], the decomposed Pt atoms would readily occupy the Ni-vacancies which earlier formed at the edges by migration [15], thereby exhibiting the radial thickening of Pt–Ni @ Pt frame with progressing time. The nonsusceptibility of Pt to dimethylglyoxime solution [16] and strong Pt–Pt bonds allowed Pt-protected frames to be retained as far as possible after the etching. These were in agreement with the experimental observations. Therefore, it is confirmed that the existence of open {100} surfaces and

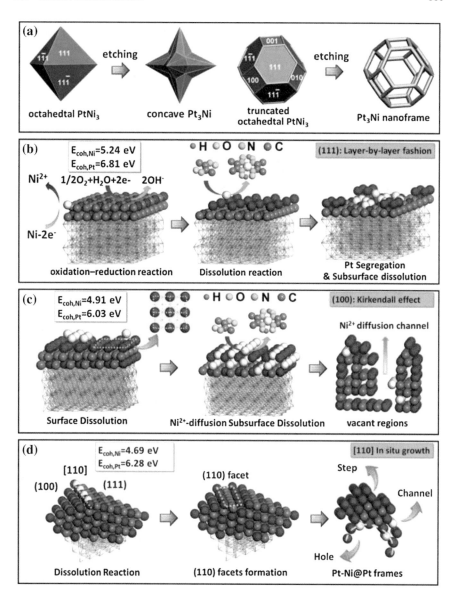

Fig. 5.4 Schematic illustration of the proposed etching mechanism of truncated octahedral PtNi$_3$ NPs. **a** The schematic illustratuion, **b** and **c** Etching processes of {111} and {100} surfaces, **d** Formation of [110] frame. Reprinted with the permission from Ref. [11]. Copyright 2014 American Chemical Society

Kirkendall effects stemming from distinct diffusion coefficients of Ni and Pt are the major reasons truncated octahedral PtNi$_3$ could yield the stable framework by chemical etching. Moreover, this mechanism can also explain the prior research of Chen et al. [9]. This kind of corrosion mechanism can be popularized to a series of other alloy systems provided that they satisfy certain conditions: (1) there are excessive active components compared with the inert components, such as PtCu$_3$; (2) there must be an active surface in the alloy polyhedral structure which follows different ways of accumulation of the components between layers, so as to guarantee the generation of the differences and internal holes during the corrosion process, such as the face-centered cubic structure formed by combining the surfaces of {110} [9], {100}–{111}, and {110}–{111}, or the body-centered cubic structure formed by combining the surface of {110} and {100}; (3) a selective chemical corrosion environment which guarantees the corresponding rearrangement of the inert components as well as the corrosion of the active components. With the corrosion mechanism taken into account, we believe that the chemical etching method will become a general means by which plenty of diversified alloy frame structures can be synthesized.

5.3.3 The Synthesis of Au-Modified Pt–Ni Trimetallic Nanoframes

To gain comprehensive insights into the structure–activity relationship of a catalyst, we designed an experiment simultaneously involving the effects of morphology and composition within a single sample. We also chose truncated octahedral PtNi$_3$ particles as the starting polyhedrons for depositing Au islands on Pt–Ni nanoframe hybrid structures by a sequential process. The structural evolution from truncated octahedral PtNi$_3$, to Au on PtNi$_3$, to Au on PtNi, and finally to Au on Pt$_3$Ni is illustrated in Fig. 5.5.

The first step of this process starts with a mild galvanic replacement between Au(III) and Ni(0). Thus, it is not surprising that Au atoms preferentially deposited on corners and Au island has been in situ formed on corners, because corner atoms have higher surface free energy than edges atoms and surface atoms. To further elucidate the nature of this unique hybrid structure, we also carried out EDS elemental mapping measurements. Unlike the homogeneous distribution of Pt and Ni throughout each particle, the turquoise dots in Fig. 5.5 clearly indicate the existence of isolated Au islands. The second step after replacement was the chemical etching process. By controlling the amount of etchant, the compositions and structures of the final obtained trimetallic nanoframes can be regulated. The compositions of these hybrid structures continued to evolve from 10 % Au on PtNi$_3$, to 10 % Au on PtNi, to 10 % Au on Pt$_3$Ni, which was consistent with the morphological transformation of the Pt–Ni cores from truncated octahedrons to truncated octahedral nanoframes. Figure 5.5 shows TEM images of the samples at three different stages, which demonstrated a morphological yield of up to 95 %. Thereby, it demonstrated that this method is very effective and versatile. It is noteworthy that the amount of Au deposited could also be controlled by simply varying the volume of the HAuCl$_4$ solution. With an increase in the amount of Au, Au island could grow and it lead to that Au clusters could not diffuse into the cores (Figs. 5.6, 5.7 and 5.10).

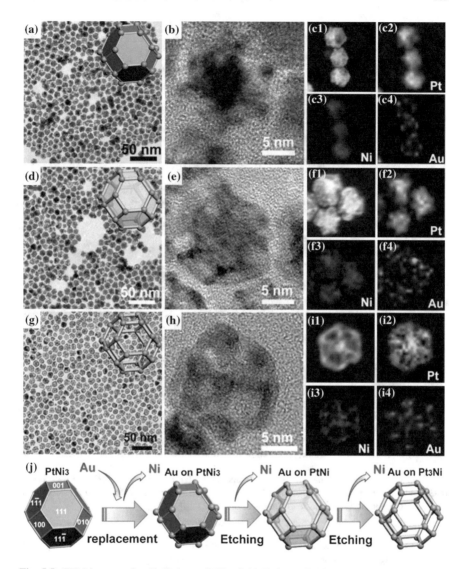

Fig. 5.5 TEM images of **a** 10 % Au on PtNi₃, **d** 10 % Au on PtNi, and **g** 10 % Au on Pt₃Ni. HRTEM images of **b** 10 % Au on PtNi₃, **e** 10 % Au on PtNi, and **h** 10 % Au on Pt₃Ni. EDS elemental mapping images of **c** 10 % Au on PtNi₃, **f** 10 % Au on PtNi, and **i** 10 % Au on Pt₃Ni. **j** Schematic illustration of the structural change from the truncated octahedral PtNi₃ to trimetallic hybrid Au islands on Pt₃Ni nanoframe. Reprinted with the permission from Ref. [11]. Copyright 2014 American Chemical Society

By simply replacing HAuCl₄ with Na₂PdCl₄ or Cu(NO₃)₂, it was possible to extend this method to construct Pd or Cu islands on Pt–Ni trimetallic nanoframes (Figs. 5.8 and 5.9).

The characterization of the spherical-aberration-electron microscopy further verifies that the lattice fringe on the corners is attributed to the {111} facet of Au,

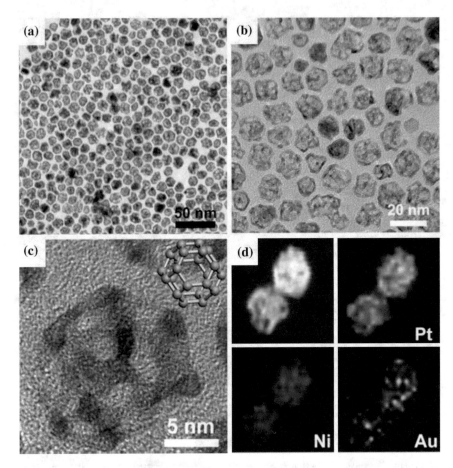

Fig. 5.6 **a** TEM image, **b** magnified TEM image, and **c** HRTEM images of 20 % Au on Pt₃Ni.
d EDS elemental mapping images of 20 % Au on Pt₃Ni. Reprinted with the permission from
Ref. [11]. Copyright 2014 American Chemical Society

while the crystal structure on the edge is consistent with the Pt₃Ni crystal phase.
The previous reports have pointed out that atoms with smaller lattice spacing such
as Cu and Ni can diffuse into the interstitial void of the atoms with larger lattice
spacing such as Au or Pt to form the alloy structure [17, 18]. As compared with Pt
or Ni, Au has larger lattice spacing, the diffusion behavior is impossible. The Au
atoms are more likely to form islands in the exterior of Pt–Ni, which further form
the Au island modified trimetal frame structure of Pt–Ni.

This chapter also employed the XRD and EDS to characterize the crystalline
structure evolution process of the Au island modified trimetal frame structure. The
XRD results showed that two sets of standard peaks can be found in the trimetal
structure, which corresponded to the face-centered cubic Au and {111}, {200},
{220} facets of Pt–Ni alloy, respectively. Neither Au–Pt nor Au–Ni alloy phase
was formed in the trimetal system. After etching, the peak corresponding to the

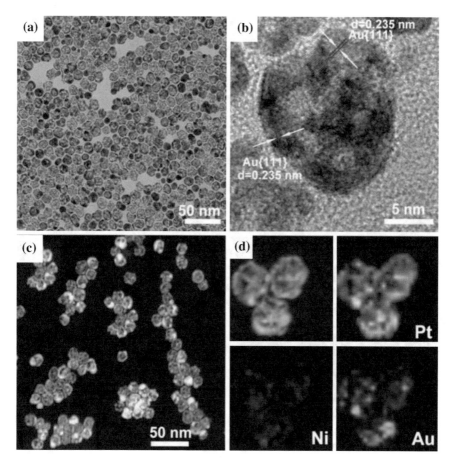

Fig. 5.7   a TEM image, **b** HRTEM image, and **c** HAADF-STEM images of 50 % Au on Pt$_3$Ni. **d** EDS elemental mapping images of 50 % Au on Pt$_3$Ni. Reprinted with the permission from Ref. [11]. Copyright 2014 American Chemical Society

Pt–Ni alloy moved towards the direction of the low angle, which increased the interplanar spacing and also demonstrated that the Ni atoms were stripped continuously from the Pt–Ni alloy phase, and the increase of the amount of Au will significantly increase the peak value corresponding to Au.

5.3.4   The Catalytic Properties of Au-Modified Pt–Ni Trimetallic Nanoframes

Recently, it was demonstrated that a third metal can indeed affect the adsorption of adsorbed substrate such as small molecule on the bimetallic surface and further affect the catalytic behavior of the catalyst [19, 20]. As shown in Fig. 5.11, the

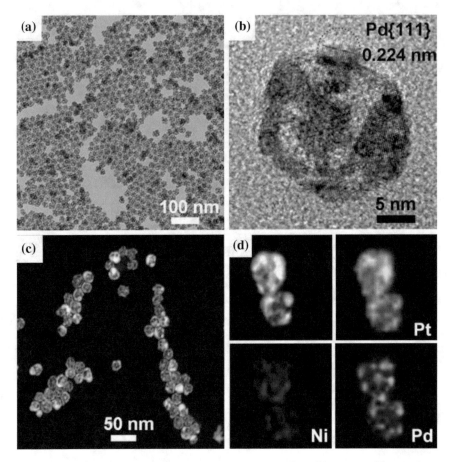

Fig. 5.8 **a** TEM image, **b** HRTEM image, and **c** HAADF-STEM images of 10 % Pd on Pt₃Ni. **d** EDS elemental mapping images of 10 % Pd on Pt₃Ni Reprinted with the permission from Ref. [11]. Copyright 2014 American Chemical Society

Pt $4f_{7/2}$ and $4f_{5/2}$ peaks of the Pt₃Ni frames corresponded to Pt^0. Similarly, the Ni $2p_{3/2}$ peak could be assigned to Ni^0. Comparatively, an apparent decrease in binding energy was observed for Au islands on the Pt₃Ni frame structures. The results strongly indicate that the Au islands on the Pt–Ni frame do have electron communication with Pt–Ni frames, where electrons were transferred from Au islands to Pt–Ni frames. The hydrogenation reaction of aromatic nitro groups is an important industrial process. However, when the reactant molecules consist of other reducible groups such as C=C and C=O, the nitro groups are difficult to be reduced in a selective manner [21]. Utilizing the hydrogenation of 4-nitrobenzaldehyde as a model probe reaction, we could evaluate the catalytic behavior of truncated octahedral PtNi₃ NPs and Pt₃Ni, Pt₃Ni nanoframes and hybrid Au on Pt–Ni with respect to both activity and selectivity. As shown in Fig. 5.11, due to their larger

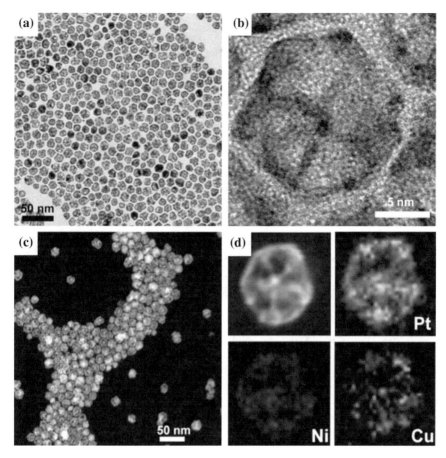

Fig. 5.9   a TEM image, **b** HRTEM image, and **c** HAADF-STEM images of 10 % Cu on Pt$_3$Ni. **d** EDS elemental mapping images of 10 % Cu on Pt$_3$Ni. Reprinted with the permission from Ref. [11]. Copyright 2014 American Chemical Society

surface area, the Pt$_3$Ni frames exhibited a higher activity turnover frequency (TOF) than the truncated octahedral PtNi$_3$ and Pt$_3$Ni particles, whereas much lower selectivity was measured (46 %). This performance is not satisfactory, and it indicates that Pt–Ni bimetallic system makes the C=C and C=O simultaneous reduction, which further showed a poor selectivity. Surprisingly, This article found that the Au on Pt–Ni showed a greatly high initial activity TOF and almost 100 % selectivity, with small amounts of Au islands decorated on the Pt$_3$Ni frames. This increase might be attributed to localized electron communication of Au atom and taking advantage of reactive sites. However, an increase in the amount of Au did not cause a further enhancement in activity but instead a significant decay. It is possible that the overgrowth of Au islands may occur on the NPs, blocking the reactive sites and thereby hindering the contact between the Au atom and Pt–Ni

Fig. 5.10 HRTEM image
(equipped with a spherical
aberration (Cs) corrector) of
10 % Au on Pt₃Ni Reprinted
with the permission from
Ref. [11]. Copyright 2014
American Chemical Society

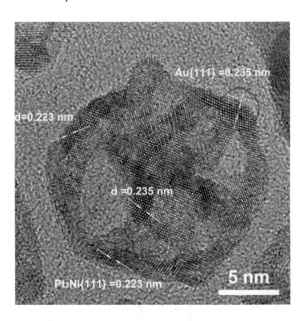

nanoframe. Thus, the introduction of a third metal may shed light on the design of novel trimetallic nanoframes exhibiting simultaneous enhancement in both activity and selectivity of catalyst.

Furthermore, we used the optimal amount of Au on the Pt₃Ni catalysts (10 %) to further study the particles' catalytic properties towards the electro-oxidation of methanol. By carving the internal Ni structure and leaving a Pt₃Ni nanoframe, chemical etching gives rise to advantageous material properties such as a desirable active surface area due to the three-dimensional edges, steps, and corners of the particle structure. There is a series of the cyclic voltammograms of the hydrogen adsorption/desorption process in Fig. 5.12. The active area of unit mass of each catalyst can be calculated by the calculation of integral. As shown in Fig. 5.12, the activity per unit mass of the frame structure is about 4 times higher than the polyhedron, thus, which showed that the nanoframes have a very excellent specific surface area. Remarkably, the doping of Au islands on the Pt₃Ni nanoframes could significantly increase the If/Ib ratio (If and Ib are the forward and backward current densities, respectively). This result indicates that the presence of Au island can facilitate the oxidation of methanol via a more effective route and prevent the generation of poisoning species such as CO [22]. To confirm this inference, we further carried out CO stripping experiments. As clearly shown in Fig. 5.11, 10 % Au on the Pt₃Ni nanoframes allowed for a higher CO-resistant activity compared with that of other catalysts. In addition, this hybrid trimetallic nanoframe exhibited outstanding electrochemical durability. As is known to all, because of the dissociation of Pt and the adsorption of some poisoning species such as CO, the conventional catalyst is often inactivated when used [23]. From Fig. 5.11, high catalytic activity was retained with this trimetallic nanoframes even after 3000 cycles,

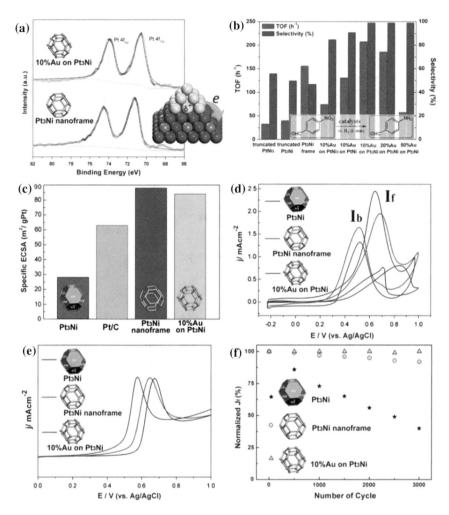

Fig. 5.11 a Pt 4f XPS spectra of trimetallic nanoframe. **b** The catalytic properties of trimetallic Nanoframes of 4-nitrobenzaldehyde hydrogenation. **c** Specific ECSAs for different catalysts. **d** Cyclic voltammograms of methanol oxidation. **e** Cyclic voltammograms of CO stripping on different catalysts. **f** The durability of electro-catalysis of methanol oxidation. Reprinted with the permission from Ref. [11]. Copyright 2014 American Chemical Society

In addition, this hybrid trimetallic nanoframe exhibited outstanding durability. As mentioned previously, the skin surface of the nanoframes obtained by chemical etching is rich in Pt, which should strengthen the stability of this structure and prevent the dissolution of Pt [24]. Moreover, the synergistic effect between the Au islands and Pt–Ni frame should weaken the binding of adsorbed and poisonous intermediates on the particle surface, thus facilitating the durability of this ternary hybrid nanoframe catalyst [25].

Fig. 5.12 Cyclic voltammograms of the trimetallic nanoframe catalyst in 0.1 M HClO₄. Reprinted with the permission from Ref. [11]. Copyright 2014 American Chemical Society

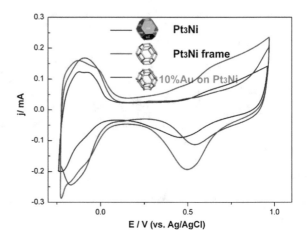

5.4 Conclusions

In this chapter, a novel chemical etching method was successfully applied to synthesize the Pt₃Ni nanoframe. The underlying mechanism for the formation of the nanoframes has been investigated thoroughly based on the theoretical study and supported by experimental observations. Moreover, by taking advantage of the high activity of the corner atoms of the polyhedron, we successfully constructed a novel trimetallic nanoframe by sequential strategies. This specific structure not only preserves most of the active sites, but also can exhibit the characteristics of a multicenter, multifunctional multiple heterostructure. Finally, comparative model experiments proved that we developed an ideal trimetallic nanoframe catalyst, which can modulate simultaneous activity and selectivity. This work has important significance for further understanding the structure–activity relationship of catalysts and designing more efficient and practical catalysts.

References

1. Liu Z-P, Hu P (2003) General rules for predicting where a catalytic reaction should occur on metal surfaces: a density functional theory study of CH and CO bond breaking/making on flat, stepped, and kinked metal surfaces. J Am Chem Soc 125(7):1958–1967
2. Lebedeva N, Koper M, Feliu J, Van Santen R (2002) Role of crystalline defects in electrocatalysis: mechanism and kinetics of CO adlayer oxidation on stepped platinum electrodes. J Phys Chem B 106(50):12938–12947
3. Quan Z, Wang Y, Fang J (2012) High-index faceted noble metal nanocrystals. Acc Chem Res 46(2):191–202
4. Cui C, Gan L, Heggen M, Rudi S, Strasser P (2013) Compositional segregation in shaped Pt alloy nanoparticles and their structural behaviour during electrocatalysis. Nat Mater 12(8):765–771
5. Sun Y, Xia Y (2002) Shape-controlled synthesis of gold and silver nanoparticles. Science 298(5601):2176–2179

6. McEachran M, Keogh D, Pietrobon B, Cathcart N, Gourevich I, Coombs N, Kitaev V (2011) Ultrathin gold nanoframes through surfactant-free templating of faceted pentagonal silver nanoparticles. J Am Chem Soc 133(21):8066–8069

7. Hong X, Wang D, Cai S, Rong H, Li Y (2012) Single-crystalline octahedral Au–Ag nanoframes. J Am Chem Soc 134(44):18165–18168

8. Xia BY, Wu HB, Wang X, Lou XW (2012) One-pot synthesis of cubic PtCu3 nanocages with enhanced electrocatalytic activity for the methanol oxidation reaction. J Am Chem Soc 134(34):13934–13937

9. Chen C, Kang Y, Huo Z, Zhu Z, Huang W, Xin HL, Snyder JD, Li D, Herron JA, Mavrikakis M, Chi M, More KL, Li Y, Markovic NM, Somorjai GA, Yang P, Stamenkovic VR (2014) Highly crystalline multimetallic nanoframes with three-dimensional electrocatalytic surfaces. Science 343(6177):1339–1343

10. Wu Y, Wang D, Niu Z, Chen P, Zhou G, Li Y (2012) A strategy for designing a concave Pt–Ni alloy through controllable chemical etching. Angew Chem Int Ed 124(50):12692–12696

11. Wu Y, Wang D, Zhou G, Yu R, Chen C, Li Y (2014) Sophisticated construction of Au islands on Pt–Ni: an ideal trimetallic nanoframe catalyst. J Am Chem Soc 136(33):11594–11597

12. Yin Y, Rioux RM, Erdonmez CK, Hughes S, Somorjai GA, Alivisatos AP (2004) Formation of hollow nanocrystals through the nanoscale Kirkendall effect. Science 304(5671):711–714

13. González E, Arbiol J, Puntes VF (2011) Carving at the nanoscale: sequential galvanic exchange and Kirkendall growth at room temperature. Science 334(6061):1377–1380

14. Wu Y, Wang D, Chen X, Zhou G, Yu R, Li Y (2013) Defect-dominated shape recovery of nanocrystals: a new strategy for trimetallic catalysts. J Am Chem Soc 135(33):12220–12223

15. Kellogg G, Feibelman PJ (1990) Surface self-diffusion on Pt (001) by an atomic exchange mechanism. Phys Rev Lett 64(26):3143

16. Gazda DB, Fritz JS, Porter MD (2004) Determination of nickel (II) as the nickel dimethylglyoxime complex using colorimetric solid phase extraction. Anal Chim Acta 508(1):53–59

17. Chen W, Yu R, Li L, Wang A, Peng Q, Li Y (2010) A seed-based diffusion route to monodisperse intermetallic CuAu nanocrystals. Angew Chem 122(16):2979–2983

18. Niu Z, Wang D, Yu R, Peng Q, Li Y (2012) Highly branched Pt–Ni nanocrystals enclosed by stepped surface. Chem Sci 3:1925–1929

19. Zhang S, Guo S, Zhu H, Su D, Sun S (2012) Structure-induced enhancement in electrooxidation of trimetallic FePtAu nanoparticles. J Am Chem Soc 134:5060–5063

20. Wang DY, Chou HL, Lin YC, Lai FJ, Chen CH, Lee JF, Hwang BJ, Chen CC (2012) Simple replacement reaction for the preparation of ternary Fe1–xPtRux nanocrystals with superior catalytic activity in methanol oxidation reaction. J Am Chem Soc 134(24):10011–10020

21. Blaser HU, Steiner H, Studer M (2009) Selective catalytic hydrogenation of functionalized nitroarenes: an update. Chemcatchem 1(2):210–221

22. Xu D, Liu Z, Yang H, Liu Q, Zhang J, Fang J, Zou S, Sun K (2009) Solution-based evolution and enhanced methanol oxidation activity of monodisperse platinum-copper nanocubes. Angew Chem Int Ed 48(23):4217–4221

23. Tang L, Han B, Persson K, Friesen C, He T, Sieradzki K, Ceder G (2009) Electrochemical stability of nanometer-scale Pt particles in acidic environments. J Am Chem Soc 132(2):596–600

24. Matanovic I, Garzon FH, Henson NJ (2011) Theoretical study of electrochemical processes on Pt–Ni alloys. J Phys Chem C 115(21):10640–10650

25. Stamenkovic VR, Mun BS, Arenz M, Mayrhofer KJ, Lucas CA, Wang G, Ross PN, Markovic NM (2007) Trends in electrocatalysis on extended and nanoscale Pt-bimetallic alloy surfaces. Nat Mater 6(3):241–247